2750 BCE

1500 BCE

100 CE

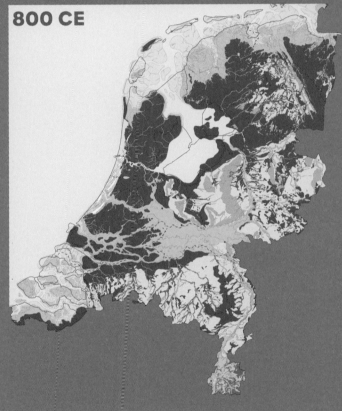

800 CE

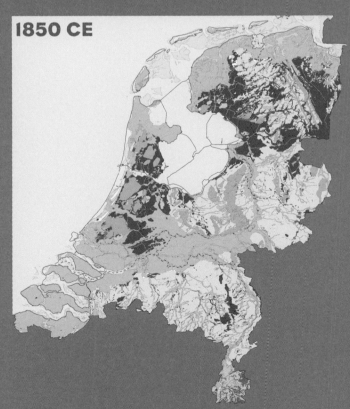

1850 CE

2000 CE

ATLAS
OF THE
HOLOCENE
NETHERLANDS

ATLAS
OF THE
HOLOCENE
NETHERLANDS

LANDSCAPE AND HABITATION SINCE THE LAST ICE AGE

Edited by **Peter Vos**, **Michiel van der Meulen**, **Henk Weerts** and **Jos Bazelmans**

AMSTERDAM UNIVERSITY PRESS

Editors
Peter Vos, Michiel van der Meulen, Henk Weerts and Jos Bazelmans

Project management
Jos Bazelmans

Map compilation
Peter Vos and Sieb de Vries

Final editing
Jos Bazelmans and Eelco Beukers

Palaeogeographical map design
Han Bruinenberg, Marjolein Haars, Veronique Marges and Sieb de Vries

Figure design
Marjolein Haars, Menne Kosian, Mikko Kriek, Klaas van der Veen and Sieb de Vries

Text
Palaeogeography
Peter Vos, Henk Weerts, Jos Bazelmans, Bob Hoogendoorn and Michiel van de Meulen

Archaeology
Stijn Arnoldussen, Henk Baas, Jos Bazelmans, Jan van Doesburg, Bert Groenewoudt, Tessa de Groot, Jan Willem de Kort, Hans Peeters, Lammert Prins, Michel Lascaris, Theo Spek, Eelco Rensink, Liesbeth Theunissen and Henk Weerts

Palaeobotany and archaeozoology
Otto Brinkkemper and Roel Lauwerier

Glossary
Jos Bazelmans, Michiel van der Meulen and Henk Weerts

Map consultants
Henk Baas, Jos Bazelmans, Roy van Beek, Meindert van den Berg, Ingwer Bos, Otto Brinkkemper, Kim Cohen, Jan van Doesburg, Bert Groenewoudt, Tessa de Groot, Mark Hijma, Hans Peeters, Lammert Prins, Eelco Rensink, Ad van der Spek, Liesbeth Theunissen and Henk Weerts

Edition
This atlas was originally published in 2011 in Dutch under the title *Atlas van Nederland in het Holoceen*, Amsterdam, Prometheus. In 2018 the 9th revised edition was published. The revision and this English edition were made possible by a financial contribution from the 'Mapping archaeological knowledge' programme of the Cultural Heritage Agency of the Netherlands.

Translation Annette Visser
Cover illustration Peter Vos and Sieb de Vries
Book design Suzan Beijer

Originally published in 2018 by Uitgeverij Prometheus, Amsterdam, as: *Atlas van Nederland in het Holoceen*, by P. Vos, J.G.A. Bazelmans, M. van der Meulen, H. Weerts. Copyright © 2018 by P. Vos, J.G.A. Bazelmans, M. van der Meulen, H. Weerts

First English edition 2020
© The copyright for the maps and individual texts rests with the Cultural Heritage Agency of the Netherlands, TNO and Deltares.

ISBN   978 94 6372 443 2
DOI   10.5117/9789463724432
NUR   933

© P. Vos, M. Van der Meulen, H. Weerts, J. Bazelmans / Amsterdam University Press B.V., Amsterdam 2020

CONTENTS

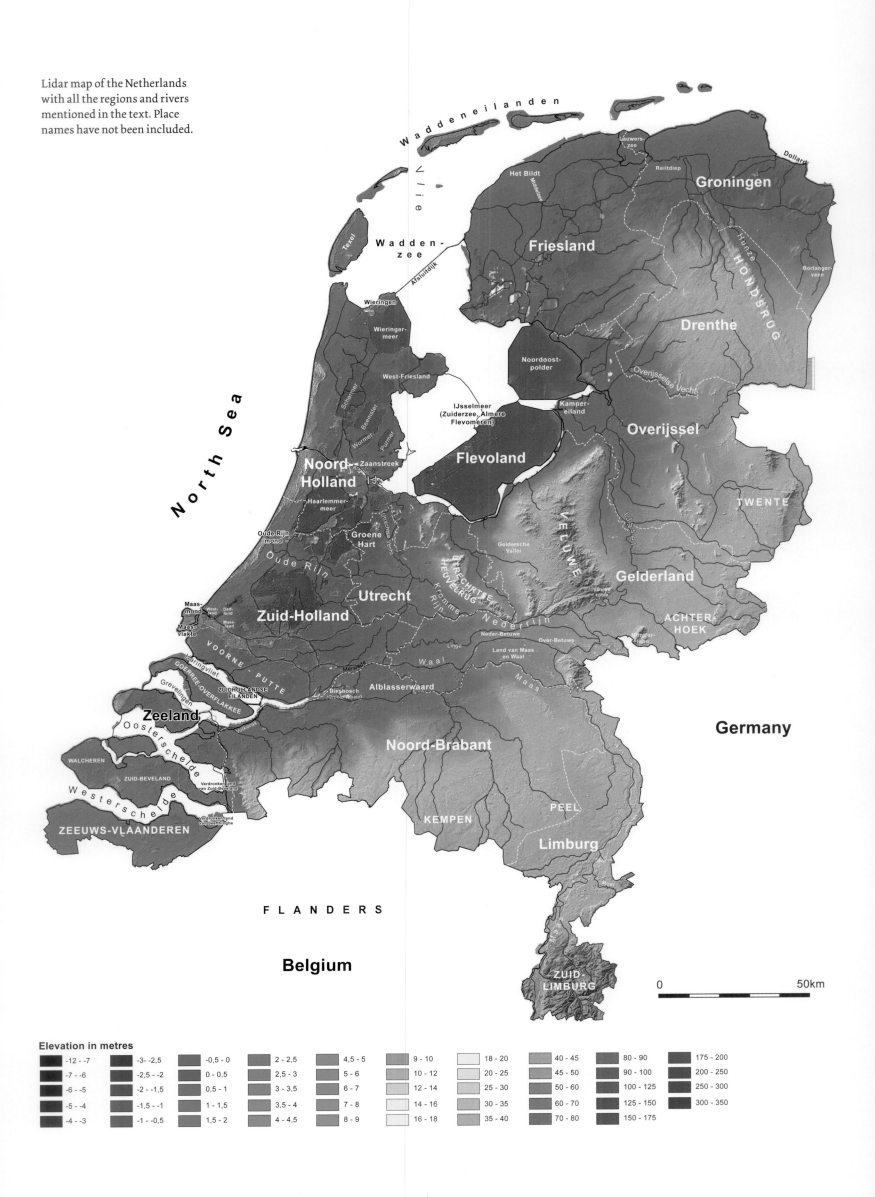

Lidar map of the Netherlands with all the regions and rivers mentioned in the text. Place names have not been included.

Waddeneilanden

North Sea

Wadden-zee

Texel

Vlie

Afsluitdijk

Wieringen

Wieringer-meer

West-Friesland

Schermer

Beemster

Wormer

Purmer

Zaanstreek

Noord-Holland

IJ

Haarlemmer-meer

Oude Rijn mond

Oude Rijn

Groene Hart

Utrechtse Vecht

Utrecht

Kromme Rijn

UTRECHTSE HEUVELRUG

Maas-mond

West-land

Delf-land

Maas-land

Zuid-Holland

Maas-vlakte

VOORNE - PUTTE

Haringvliet

GOEREE-OVERFLAKKEE

Grevelingen

ZUIDHOLLANDSE EILANDEN

Biesbosch (Groene Waard)

Merwede

Alblasserwaard

Volkerak

Zeeland

Oosterschelde

WALCHEREN

ZUID-BEVELAND

Verdronken land van Zuid-Beveland

Westerschelde

Verdronken land v.h. Swninghe

ZEEUWS-VLAANDEREN

Het Bildt

Middelzee

Friesland

Lauwers-zee

Reiitdiep

Dollard

Groningen

Hunze

HONDSRUG

Drenthe

Bortanger-veen

Noordoost-polder

Overijsselse Vecht

Kamper-eiland

Overijssel

IJsselmeer (Zuiderzee, Almere Flevomeren)

Flevoland

TWENTE

IJssel

Gelderse Vallei

VELUWE

Veluwe-zoom

Gelderland

ACHTER-HOEK

Neder-Betuwe

Over-Betuwe

Kamper-land

Linge

Land van Maas en Waal

Nederrijn

Waal

Maas

Noord-Brabant

Germany

KEMPEN

PEEL

Limburg

Roer

FLANDERS

Belgium

Maas

ZUID-LIMBURG

0   50km

Elevation in metres

-12 - -7	-3- -2,5	-0,5 - 0	2 - 2,5	4,5 - 5	9 - 10	18 - 20	40 - 45	80 - 90	175 - 200
-7 - -6	-2,5 - -2	0 - 0,5	2,5 - 3	5 - 6	10 - 12	20 - 25	45 - 50	90 - 100	200 - 250
-6 - -5	-2 - -1,5	0,5 - 1	3 - 3,5	6 - 7	12 - 14	25 - 30	50 - 60	100 - 125	250 - 300
-5 - -4	-1,5 - -1	1 - 1,5	3,5 - 4	7 - 8	14 - 16	30 - 35	60 - 70	125 - 150	300 - 350
-4 - -3	-1 - -0,5	1,5 - 2	4 - 4,5	8 - 9	16 - 18	35 - 40	70 - 80	150 - 175	

PREFACE

This Atlas presents thirteen reconstructions of the Dutch landscape and its habitation since the last ice age. They span a period that we refer to as the Holocene. The maps and accompanying texts provide a vivid illustration of how the Netherlands has changed beyond recognition in 11,700 years in response to sea-level rise, the transport of sand and clay from the sea and large rivers, peat expansion and – not least – the growing impact of humans.

Shortly before the start of the Holocene, the Netherlands was an indistinguishable part of a vast, cold and sparsely populated tundra that extended far into the present-day North Sea. Today, it is a densely populated country, almost entirely cultivated, and protected against threats from the sea and river waters by a complex system of dykes and embankments. It is hard to imagine a greater contrast. The *Atlas* covers a fascinating period: as the maps show, the landscape has long felt the impact of climate change, sea-level rise and human intervention. The history mapped out here does not of course offer any ready-made solutions to these pressing modern-day issues. It does, however, show the intended and unintended (sometimes disastrous) consequences of human intervention in the landscape and how people have sometimes succeeded, and sometimes failed, to overcome them.

The Netherlands has a rich tradition of map-making, from Blaeu's *Atlas Maior* to the *Bosatlas*. What is so unique about this atlas is the long timespan that it covers and the way it combines knowledge about human development and about our environment. We know a good deal about the origins, evolution and habitation of the Netherlands since the last ice age – the result of generations of work. Staff from Deltares, the TNO Geological Survey of the Netherlands and the Cultural Heritage Agency of the Netherlands (RCE) have pooled their knowledge to compile an atlas that is accessible both to the interested reader and to students at secondary schools and universities. This is in keeping with the aim of these organisations to effectively document the soil-related, geological, geomorphological, archaeological and cultural heritage characteristics of our subsoil and in so doing create the conditions whereby these values can be sustainably preserved.

Almost ten thousand copies of the Dutch edition of this atlas have sold since 2011. In line with the expectation expressed in the preface to that first edition, it has played a major role in recent years in discussions about what we still do not know about the evolution of the Dutch landscape. New insights have prompted the publication of a revised and expanded edition in 2018. This first English edition is an unaltered version of the 2018 edition.

We would like to thank everyone who has made the production of this atlas possible.

The editors

INTRODUCTION

FIG. I A map by Zagwijn from 1986, at its original published size: The Netherlands in the Late Atlantic (4100 BCE). Compare the new map for 3850 BCE on p. 45, with its conspicuously greater detail, the result of data and knowledge acquired over the past 30 years.

Legend

- open water
- tidal flats
- salt marshes and clay layers
- former salt marshes and clay layers
- river deposits
- local peat formation
- peat bog
- bog
- coastal dunes and beaches
- river dunes
- land dunes
- Pleistocene
- not included
- boundary still present
- assumed boundary

0   75km

1 THE NETHERLANDS IN THE HOLOCENE

We don't often stop to think about it, but our landscape in the Netherlands has changed. Where we walk today, sand once drifted across a cold polar desert, rivers flowed between lakes and swamps, primeval forests stretched across vast expanses and the sea surged. The present-day Netherlands was formed over many thousands of years, shaped by sea currents, rivers, wind, land ice, flora and fauna. And, of course, humans have also contributed to the landscape in its present form. In past centuries we humans have become an important 'geological' force: in today's Netherlands it is almost impossible to point to an area where people *haven't* left their mark.

This book explains how the Dutch landscape has changed since the end of the last ice age, 11,700 years ago. This climate switch marked the beginning of the Holocene, the relatively warm geological epoch in which we are currently living. It was from this time on that the Netherlands slowly began to acquire the form that it has today.

To clearly illustrate how the present-day Netherlands evolved, Peter Vos and Sieb de Vries (TNO and Deltares) have created a set of thirteen palaeogeographical maps for this book. Each of these maps, which depict past geographical situations, presents a moment in time: they show how the different landscapes were distributed across the Netherlands. The maps are based on findings from tens of thousands of corings, plus a good deal of supplementary research on matters such as the age and formation of soil layers. The palaeogeographical maps lie at the heart of this atlas.

This is not the first time that a series of maps of the Holocene Netherlands has appeared. In 1986 W.H. Zagwijn of the National Geological Survey was the first to publish a similar series, albeit a rather basic one (fig. 1). Since that time a vast amount of geological data has been collected, which has helped to change our understanding of the formation of the Netherlands. This new information has been incorporated into another map series in *De ondergrond van Nederland* [The subsoil of the Netherlands] (2003). A series of eleven, much more detailed maps was created for the Cultural Heritage Agency of the Netherlands as part of *De nationale onderzoeksagenda archeologie* [The National Research Agenda for Archaeology] (2006) (version 1.0). An improved series appeared in 2011 in the first Dutch edition of this atlas (version 1.1), the result of a collaboration between Deltares, the TNO Geological Survey of the Netherlands and the Cultural Heritage Agency. Its publication marked the definitive replacement of Zagwijn's atlas as a standard reference work more than twenty years after its first appearance. In 2013 the first edition of the atlas was thoroughly revised as part of the Cultural Heritage Agency's 'Mapping archaeological knowledge' project (version 2.0). A comprehensive account of the background to the second revised edition can be found in Peter Vos'

PhD thesis, *Origin of the Dutch coastal landscape* (Utrecht, 2015). In this ninth printing of the *Atlas of the Holocene Netherlands*, all the maps have been revised once again in the light of the latest research findings (version 2.1), and two new maps have been added: for 250 BCE and 1250 CE.

Thanks to the wealth of information, the new maps are now different: they are more detailed than ever and represent a leap forward in relation to earlier reconstructions. The increased level of detail still has its limitations, however. In general, the maps in this atlas present a faithful picture at the regional level, but there are still uncertainties at the local level. For example, the landscapes and landscape changes shown in the maps don't allow us to make exact predictions pronouncements about the archaeological values we can expect at the local level. On the other hand, the maps can be useful when it comes to drawing up a research agenda. We now have much greater clarity about where we should conduct research, and what kind of research, in order to remove uncertainties.

This book differs in another respect from its predecessor. More so than in 1986, it looks at the central place that humans have occupied in the landscape. We address questions like: What kind of environment did our ancestors inhabit? How did they utilise the opportunities that the landscape offered? How did they respond to changes in the landscape? And how did they mould the landscape to suit their own purposes? The interaction between people and landscape is one of the main research themes within contemporary Dutch archaeology. This atlas therefore contains not only palaeogeographical information, as in 1986, but also archaeological and historical data. We have also added a reconstruction for the nineteenth century in order to better illustrate the role played by people, and a map for 2000 to make a comparison with today's situation easier.

The notes on the palaeogeographical maps begin each time with the natural forces that have shaped the landscape and then go on to discuss the role of humans. Human activity is explained in a case study, a story about a specific settlement that is representative of the period in question.

Several natural processes keep recurring in the notes on the maps: the relative rise in sea level, tides and wave action, the changing courses of rivers and the formation of large expanses of peat. These processes are discussed at length in the Introduction, within the context of climate change. Together with human intervention, they explain how the Dutch landscape was constantly changing and how it slowly acquired its present-day form.

Because the text contains quite a number of terms that won't be immediately familiar to everyone, a glossary is included in the back of the atlas.

2 WHAT CAME BEFORE?

Geological epoch	Age in millions of years	Sea level (2000 CE = 0)

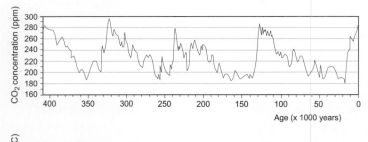

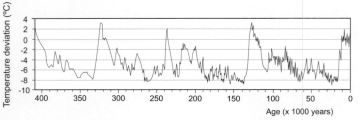

This book is about the Holocene, the geological epoch in which we currently live and which began 11,700 years ago. While that might seem a long time ago, if we think in terms of a geological timescale, it isn't. It makes up just a fraction of the four and a half billion years that the Earth has existed.

Our main reason for focusing on the Holocene is that this was the period in which the present-day Dutch landscape was largely shaped, as we will explain below. But first we need to look briefly at the preceding period in order to place the changes that occurred in the Holocene in a broader context. Looking at the past several million years, we see a highly variable climate, a constant alternation of very cold periods (ice ages or glacials) and warm periods (interglacials). In the northern hemisphere this alternation has been happening for some 2.6 million years. For decades it was believed (and taught in schools) that Europe experienced four ice ages during this period. New research methods, however, show that there were in fact many more: no fewer than twenty since the beginning of the Pleistocene (the epoch preceding the Holocene, 2.6 million to 11,700 thousand years ago) (fig. 2). The relatively warm Holocene is thus part of a long cycle and was preceded by an ice age (the Weichselian) lasting more than 100,000 years.

In recent decades a detailed picture has emerged of climate change during the Pleistocene. It is based on the ratio between two different kinds of oxygen atoms, or isotopes: the normal variant (^{16}O), which accounts for 99.8 percent of all isotopes, and the somewhat heavier variants (^{17}O and ^{18}O isotopes). When the climate on Earth is warmer, proportionately more heavy oxygen evaporates and the amount of ^{18}O in the atmosphere increases slightly, while dropping slightly in the oceans. When it is colder on Earth, the reverse happens. These changes are meticulously recorded in sediments on the seabed and in the ice caps of Antarctica and Greenland. Seabed sediment is largely made up of the calcium skeletons of small marine creatures. When they die, they sink to the sea floor, where they accumulate, layer upon layer. These skeletons contain oxygen, with a measurable ^{16}O/^{18}O ratio. A similar analysis is possible for corings taken from the ice caps of Antarctica and Greenland, where the oxygen isotope ratio is obtained from the frozen water. These studies of ice cores (up to a maximum of 800,000 years old) and deep-sea sediment cores (sometimes millions of years old) enable us to reconstruct the changes in temperature at that location, based on the estimated water depth and the ratio of ^{16}O to ^{18}O.

The causes of the climatic variations recorded in the deep sea and ice are complex. They result from slight fluctuations in Earth's orbit around the sun (eccentricity), as well as shifts in the tilt (obliquity) and wobble (precession) of the Earth's axis. The Serbian geophysicist and mathematician Milutin Milanković calculated the size and duration of these variations (now known as the Milanković cycles). There is a growing body of geological evidence to support his theory – now generally accepted – that these cosmic cycles of 23,000, 41,000 and 100,000 years cause climate changes on Earth. The cycles have a particular influence on temperatures in summer (cool or warm summers) and winter (cold or very cold winters). Seasonal differences in mid-latitudes are also significant under some circumstances and minor under others. These variations cause fluctuations in the Earth's climate.

There also seems to be a link between global temperature and fluctuations in the carbon dioxide content of the atmosphere. Periods with a low concentration of atmospheric carbon dioxide (CO_2) have coincided with ice ages, and periods of high concentration with warm periods (fig. 3). CO_2 is a greenhouse gas: if the atmosphere contains more CO_2, it retains more heat and global temperatures climb. This variation in the concentration of atmospheric CO_2 has been happening since the dawn of time. Since the nineteenth century, however, humans – by burning large quantities of 'ancient' carbon from fossil fuels such as coal, oil and natural gas – have for the first time influenced the composition of the atmosphere and therefore probably along with it the climate on Earth.

A third and final natural factor affecting the Earth's climate is the position of the continents. As we all know, the continents are moving very slowly towards or away from one another – a phenomenon known as plate tectonics. The location of large land masses affects the temperature on Earth because continents around the equator receive more solar energy than ones around the poles. The Milanković cycles also have a stronger effect near the poles than at the equator. In the past tens of millions of years – these are the timescales we should be thinking in – relatively more continents have been located at the poles and the Milanković cycles therefore have had a strong effect.

Thus, the climate has switched several times during the last 2.6 million years from cold to warm and vice versa. Nor are temperatures constant within an ice age, but can vary from cold to very cold. What we do notice is that an ice age always ends with a very cold period lasting several thousand to tens of thousands of years, after which temperatures very quickly become significantly warmer (fig. 3). We know that sudden switches of this kind occurred 320,000, 240,000, 130,000 and 11,700 years ago.

It goes without saying that these climate switches had major consequences for the landscape of the time and for humans and their ancestors. The same is true of the change that occurred 11,700 years ago and which marked the beginning of the Holocene. Without a doubt the most obvious consequence was changes in flora and fauna. During the last ice age the area we now call the Netherlands wasn't covered by land ice, although it did have a tundra-like landscape with an icy climate. This all changed in a short space of time with the sudden warming at the onset of the Holocene.

This warming also set other processes in motion: the melting of the ice caps and the corresponding rise in sea levels. Once again, these changes would have a major impact on the Netherlands.

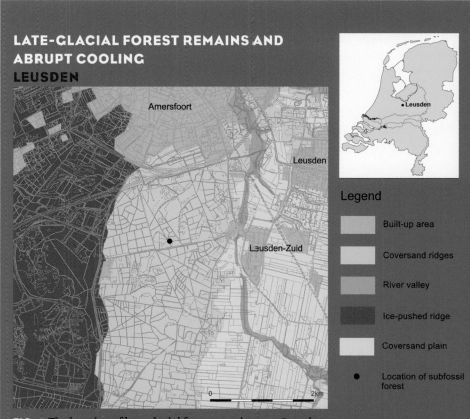

LATE-GLACIAL FOREST REMAINS AND ABRUPT COOLING
LEUSDEN

FIG. 4 The location of late-glacial forest remains near Leusden (c. 10.850 BCE)

The last ice age came to an abrupt end about 14,700 years ago. Birch, willow and juniper found their way to the Netherlands. The birch forests were later replaced by pine. Then, some 12,850 years ago, the climate in the northern hemisphere suddenly turned much colder again. This cold snap lasted for a little over one thousand years, until the beginning of the Holocene. Forests disappeared, grasses and herbs predominated and sand drifts became common.

Subfossil remains of late-glacial forests are very rare. In 2016 the remains of 170 pine trees were excavated at Leusden. Based on their tree rings and ¹⁴C-dates, they can be assigned to two groups, the oldest of which grew under warm conditions and the youngest probably falling victim to the temporary cooling.

Climate scientists are convinced that the abrupt and dramatic cooling was the result of a catastrophic outflow of enormous volumes of cold, fresh water over the relatively warm Atlantic Gulf Stream, bringing the latter to an abrupt standstill. The water had originated from a lake on the southern edge of the North American ice cap. The southward shift in the turning point of the Gulf Stream meant that the southwesterly winds that governed the weather in our region brought considerably less warmth to the northern latitudes. In these cold conditions, the gradual expansion of large quantities of sea ice southward as far as the coast of Brittany did the rest.

3 RISING SEA LEVELS

FIG. 5 North-western Europe at the end of the last ice age, in c. 10,500 and c. 8000 BCE.

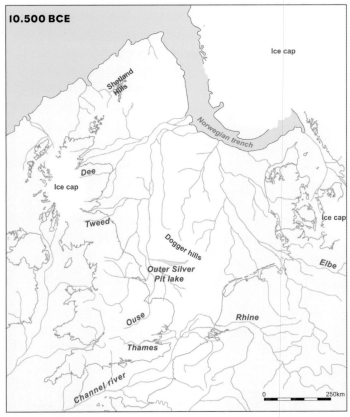

When the Holocene began 11,700 years ago, the Netherlands wasn't yet located on the coast. It was part of a vast northwestern European plain, and the British Isles were still part of the mainland of Europe (fig. 5).

The topography at that time had largely evolved during the Pleistocene: it is this Pleistocene topography (fig. 21) that literally underpins the developments described in this atlas. In the higher part of the Netherlands (roughly the southeastern half of the country) the deposits from the Pleistocene or earlier still lie almost on the surface. Even so, the landscape there has undergone significant changes. For example, extensive tracts of peatland developed, much of which has since been excavated. These changes are quite modest, however, compared with what happened in the western and northern coastal areas and in the river region. There, during the Holocene, thick layers of sand and clay were deposited on top of the Pleistocene sediments and extensive peat layers developed. The palaeogeographical maps reveal that this 'Holocene' part of the Netherlands has been the most dynamic.

The main driving force behind these changes was the rise in sea level in relation to the land. When the temperature began climbing rapidly 11,700 years ago, the lowlands filled with meltwaters from the polar ice caps within the space of 6000 years, partly submerging what is now the North Sea and even part of the present-day Netherlands.

On a geological timescale this drowning occurred very rapidly. The higher the sea level, the further eastward the coastline crept. In the flooded parts of the western and northern Netherlands, the tidal landscape of the Waddenzee came into existence, where sand and clay were deposited. This material was mainly transported from the North Sea, where the current and wave action worked sand and silt loose from the seabed and carried it to the coastal area and mudflats. This sedimentation from the sea is still evident in the sandy beach barriers along the coast. The environment was calmer in places where the water could penetrate via inlets into tidal basins behind the coast. Particles smaller than sand could also settle there and so we find clay and silt as well as sand. The Rhine and Meuse rivers also supplied sediment, but this accounted for ca 10% of the total quantity that made up the Holocene Netherlands.

Because the sea level continued to rise in relation to the land, the tidal areas shifted in an inland direction. It eventually encompassed more than a third of the current surface area of the Netherlands, including most of the current Randstad (a chain of adjacent metropolitan areas in the central western Netherlands). Even the parts of the landscape that were still above sea level were affected. As will be explained below, this is because the sea level also has an impact on river courses and the groundwater table – and hence on vegetation and peat formation.

Thus, climate changes have played a vital role in shaping the Dutch landscape. And yet climate warming and the resulting sea-level rise are just one factor. This is because, apart from minor fluctuations, average temperatures have not risen much since the temperature surge around 11,700 years ago. The melting of the ice caps in North America and Scandinavia went on for several thousand years, but came to a halt some 6000 years ago. Since then roughly the same amount of ice is formed each year as has melted.

The fact that the sea level in the Netherlands continued to rise in relation to the land (albeit it at a slower pace) now had a different cause – subsidence. There were, and continue to be, three different geological phenomena behind this subsidence: glacio-isostasy, tectonics and soil compaction (see panel). These differ from one region to another, as can clearly be seen in Figure 6. Flanders and Zeeland sank at a slower rate than the western and northern Netherlands. But there was an overall subsidence, and therefore a continuing relative rise in sea level.

Despite this, the rapid drowning of the Netherlands came to an end and the coastline stopped shifting eastward. This is because from that time on the sediment supply to the coast and peat formation kept pace with subsidence. The results can be seen in Figure 7. Without a supply of sediment, half of the Netherlands would now be under water. The fact that this didn't happen is because sufficient quantities of sea and river sediments were deposited in that half of the country, allowing peat to form. At times when sand deposition along the coast occurred more rapidly than subsidence, the coast even began to extend once more in a *westerly* direction. This was of course more likely to happen at times of modest relative sea-level rise than when sea levels were still rising rapidly.

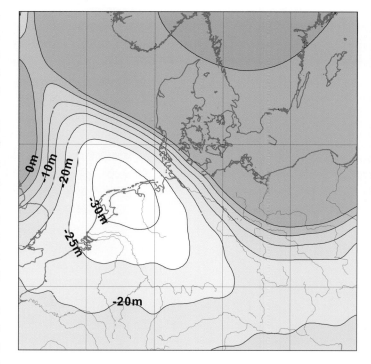

FIG. 6 Ground subsidence and uplift as a result of movements of the Earth's crust following the melting of the land ice in northwestern Europe, in metres, compared with the current height since 18,000 BCE.

As we said earlier, sediment is also transported by rivers. When rivers overflow their banks, river clay is deposited on the surrounding land or sandy levees are formed along the riverbanks. When rivers flow normally along their beds to the sea, all the sediment is transported to the sea, where it settles in the deltas and can lead to localised coastal expansion.

It is clear therefore that other factors besides relative sea-level rise have contributed to the formation of the Netherlands. One such factor is the activities of the big rivers, another the tides and wave action.

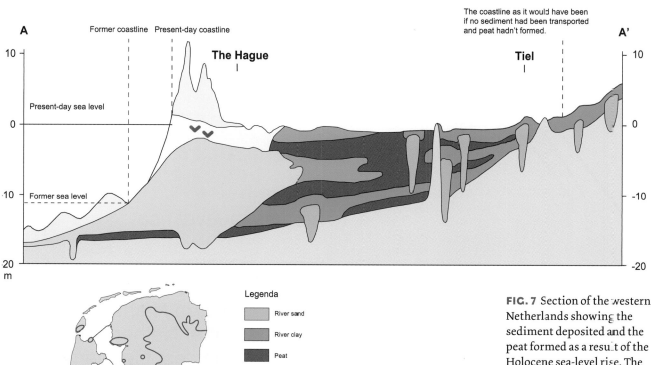

Legenda

- River sand
- River clay
- Peat
- Mudflat sand
- Marine clay
- Dune sand
- Marine sand
- Beach sand
- Pleistocene sand

The coastline as it would have been if no sediment had been transported and peat hadn't formed.

A ——— A' Profile

FIG. 7 Section of the western Netherlands showing the sediment deposited and the peat formed as a result of the Holocene sea-level rise. The diagram shows the sea level in about 6000 BCE and where today's coast would be in the absence of this sedimentation and peat formation.

Research on the course of sea-level fluctuations tends to focus on the characteristics and fossil content of sediments. For example, if seashells are found in a particular layer, we know that that area was part of the sea when these organisms were deposited. With the aid of certain fossils, we can even say something about deposition depths. Conversely, the presence of tree remains or signs of soil formation indicate that the area was land at that time. In places that lend themselves to such analysis (including the Netherlands), researchers have been able to reconstruct the Holocene sea-level rise fairly accurately by dating peat layers.

Research has shown, however, that it is not enough to carry out these measurements in a single location. This is because the changes in water depth that we observe are determined not only by in increase and reduction in the volume of water in the oceans, but also by ground movements. In other words, we measure a deepening irrespective of whether the ground subsides or the sea level rises. Thus, we are reconstructing the *relative* and not the *absolute* sea-level movement, whereby all places on Earth have their own sea-level history (fig. 8).

Local ground movements have different causes: tectonics, glacio-isostasy and soil compaction.

Plate tectonics (the gradual drifting apart and collision of tectonic plates) mean that the land rises in some places and subsides in others. This phenomenon also occurs in the Netherlands,

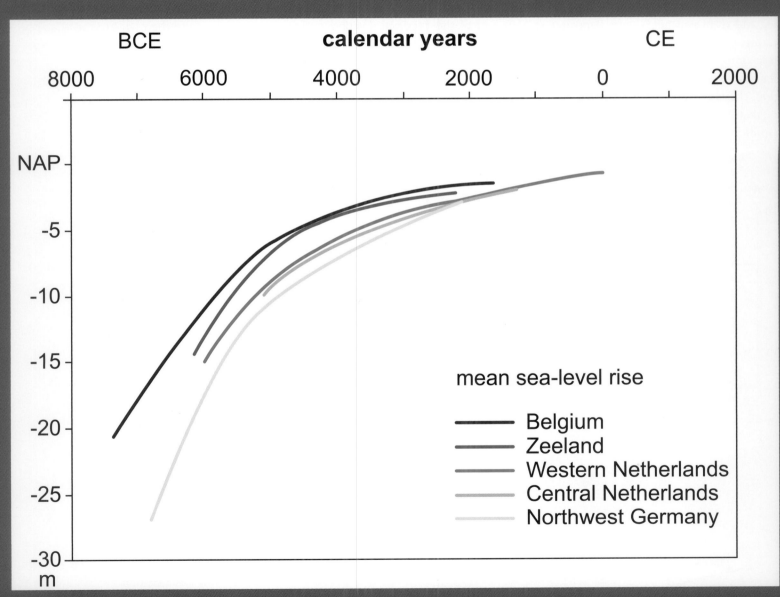

FIG. 8 The sea-level rise along the North Sea coast of in the Netherlands, Belgium and Germany, measured for the last 8000 years.

which is located on the transition of the sinking North Sea basin towards the stable European continent. Roughly speaking, the western half of the Netherlands is therefore sinking in relation to the eastern half, by about 1.5 m since the beginning of the Holocene. The presence of land ice also has an impact as the thick layer of ice pushes down on the Earth's crust below. In the asthenosphere, the relatively 'soft' part of the underlying mantle, at a depth of 80 to 300 km, material under the ice cap 'flows' away, forcing the adjacent areas up. When the ice melts, the crust beneath the ice cap springs back again, and the crust near the ice cap sinks down again. Known as glacio-isostasy, this phenomenon caused the Netherlands (which lay just beyond the land ice) to *sink* during the Holocene, when the land ice melted (fig. 9). Because the rocks deep in the Earth's mantle 'flow' very sluggishly, we are still feeling the effects of this process, even though the major shrinking of the ice cap came to a standstill 6000 years ago.

Compaction, finally, is a phenomenon of the soil itself. It too is subject to change. Fresh peat, which is 90% water, becomes more compact as a result of drainage or being overlain by younger deposits. In the most extreme case only about 10% of the original volume remains. Added to that, if peat is exposed to the air, it 'oxidises' (decays) and disappears as CO_2, among other things, into the atmosphere. Drainage played a key role in peatland reclamation and, today too, we drain peat by pumping the polders dry. Without meaning to, we humans have contributed significantly to subsidence by causing the compaction of peat soils. Clay soils can also become slightly compacted through drainage, but they don't oxidise.

After decades of research we now know that sea-level reconstructions are only valid for small areas. Fifty kilometres away, the situation may be quite different because of the specific combination of the above three factors. Nevertheless, it is generally true to say that the land surface of the Netherlands has continued to subside during the Holocene due to tectonics, glacio-isostasy and compaction.

This knowledge is very much at odds with the thinking that prevailed until

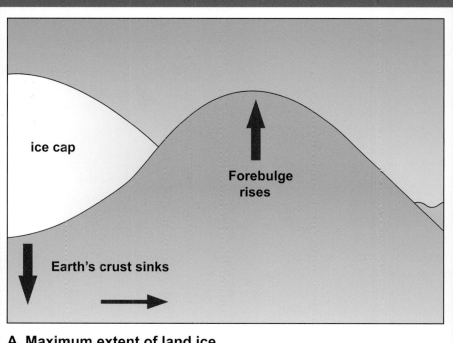

A Maximum extent of land ice

ice cap

Forebulge rises

Earth's crust sinks

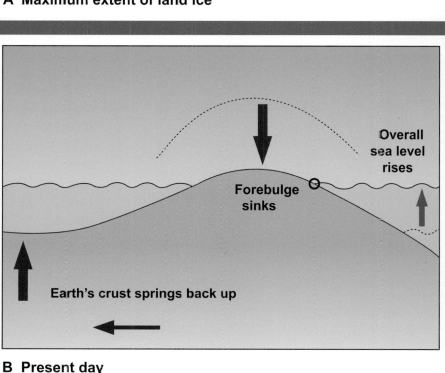

B Present day

Overall sea level rises

Forebulge sinks

Earth's crust springs back up

FIG. 9 How isostasy works. The ice cap presses the Earth's crust down across an area up to 150-180 km beyond the ice cap. Further afield is an area called the 'forebulge', which is pushed up. When the ice cap melts, the depressed crust springs back up and the forebulge sinks. In relation to land ice from the last ice age, the Netherlands is situated on this forebulge and has therefore *subsided* as a result of the melting ice.

the 1980s. Many researchers at that time believed that the sea-level rise hadn't been steady and gradual but was subject to fluctuations. They thought that periods of accelerated rise ('transgressions') alternated with periods in which the sea level rose less rapidly, or even fell ('regressions'). The geological maps of the Holocene marine deposits in the Netherlands were even based on this way of thinking. But now that we have much more data at our disposal than we did then, we know that these fluctuations probably didn't happen. Each coastal region has its own sedimentary history.

The most recent cause of land subsidence in the Netherlands is natural gas extraction and salt mining. As with peat oxidation in the polders, humans have now become the main causes of subsidence. The recent sea-level rise is also ascribed to human activity: the burning of large quantities of fossil fuels, resulting in rising CO_2 levels in the atmosphere and probably in global warming.

4 TIDES AND WAVES SHAPE THE COAST

FIG. 10 Tidal
movement in
the North Sea
and differences
in tide height
along the coast
(in cm).

NAP = Amsterdam
Ordnance Datum;
MHW = Mean High
Water;
MLW = Mean Low
Water.

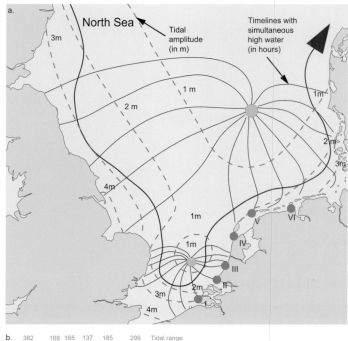

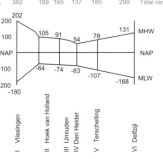

Tidal currents and wave action, which displaced large volumes of sediment, are responsible for the formation of the Dutch coast.

Twice a day, a tidal wave enters the North Sea from the Atlantic Ocean in the north. It travels around Scotland and then moves south down the east coast of England (fig. 10). Along that coast, it encounters resistance from the shallow seabed and, where the sea narrows, it is reflected. The combination of these two phenomena increases the differences in tidal height from the central part of the North Sea towards the coast. Because the effect of resistance and reflection is not the same everywhere, there is a great variation in high and low tide (tidal range) along the North Sea coast. It is tidal range that largely determined the evolution of a particular stretch of coastline.

The coastal landscape is classified into four zones, based on how often an area is flooded by the sea (fig. 11). Parts located below mean low water level belong to the 'subtidal landscape', which includes large tidal channels, sand banks that are permanently under water and tidal inlets. The 'intertidal landscape' is the area between mean low water level and mean high water level. It consists of bare, sandy tidal flats and clayey mudflats. The part of the coastal area that lies above mean high water level and is only inundated during spring tide and major storms is classified as 'supratidal landscape'. It comprises salt marshes intersected by tidal creeks and covered with salt-loving vegetation. During storms, sediment is deposited along creeks and on the edges of salt marshes facing the sea, giving rise to levees and salt-marsh ridges respectively. The area above the maximum storm surge level (or extreme high water level) is classified as 'terrestrial landscape': coastal dunes, (uncultivated) coastal peat and Pleistocene soils.

A distinction is made in the Dutch coastal landscape between tidal systems containing large rivers and ones without. A tidal system into which a major river discharges is called an

FIG. 11 A tidal
basin in a mudflat
area.

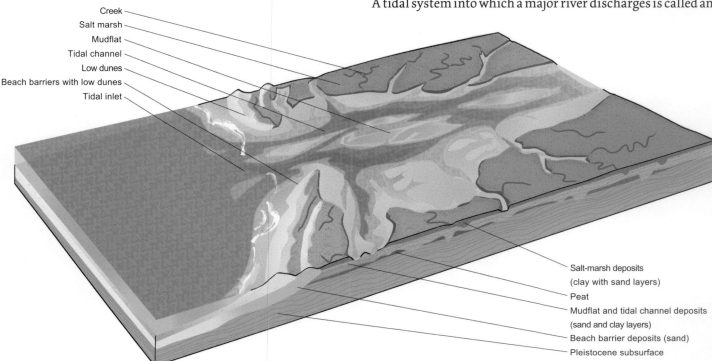

'estuary'. Estuaries fill up with sediment from the sea, along with sand and clay that is transported by the river and with peat that has formed in the floodplains. Tidal systems into which only small, local rivers discharge are called 'tidal basins'. Such basins no longer exist in the Netherlands as a result of accretion and the subsequent construction of embankments in medieval times. The Dutch coastal area also featured lagoons: large, permanently inundated areas in the coastal landscape that had an open connection to the sea. Lagoons tend to have a very modest tidal range. Before being closed off from the sea in 1932, the Zuiderzee was one such lagoon.

A tidal wave from the North Sea – like the one described above – that enters a tidal system becomes distorted. Here too, resistance and reflection play a key role. Resistance from the seafloor slows down the incoming wave. The effect is greatest where the seabed is shallow. In addition, the incoming wave is reflected by the edges of the tidal system. If the reflected wave converges with the incoming wave, it is amplified and the tidal range in the system increases. If the incoming and returning tidal waves move against one another, the incoming wave is attenuated and the tidal range declines. The shape of the tidal system determines how resistance and reflection work together. In a wide, open tidal basin, the effects of seabed resistance are greater than those of reflection from the edges of the basin. In such a system the tidal range decreases inland. In a narrow, funnel-shaped tidal system, such as an estuary, the influence of reflection is much more significant than that of resistance. As a result, the tidal range inland first increases until the river course becomes so narrow that the effect of the resistance dominates and the tidal range declines once more.

Thus, changes in the shape of tidal systems have a strong impact on tidal range and maximum tide heights. A good example is the accretion of tidal basins (fig.12). In an open tidal basin, there is a linear relationship between the size of a tidal inlet and the tidal volume of a basin system. When a tidal basin silts up, the tidal volume declines and so too does the size of the tidal inlets and channels. As a result of the silting-up, the incoming tidal wave encounters more seabed resistance and the maximum tide heights in the system decline. If accretion persists and the tidal basin is completely cut off from the sea by a beach barrier, the tide disappears altogether. This loss of natural drainage to the sea causes the groundwater level in the dried-out tidal area to rise. The area becomes wetter, peat starts to grow and a peat layer forms on top of the former tidal deposits.

Fluctuations in tide height as a result of changes in the shape of the coastal system had a major impact on habitation conditions in the coastal area. When mean and extreme high water levels fell, parts of the salt-marsh area became permanently dry and those locations could be inhabited on a permanent basis. This is what happened in West Friesland when the West Frisian tidal inlet closed definitively between 2000 and 1400 BCE. In the northern Netherlands, the Wadden sea coast remained open and the salt-marsh areas were regularly flooded during storms. The salt-marsh areas in the north were only habitable once terps were built, which occurred from the early Iron Age until the building of dykes in the High Middle Ages.

As well as tidal currents, waves play a major role in the formation of the coastal area. Waves transport sand from the shallow bed of the North Sea to the coast, which means that

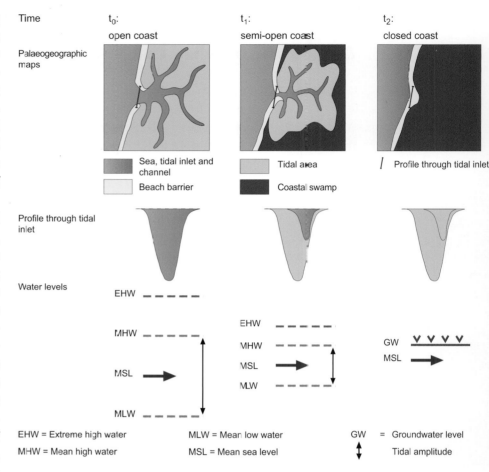

| Time | t_0: open coast | t_1: semi-open coast | t_2: closed coast |

Palaeogeographic maps

- Sea, tidal inlet and channel
- Beach barrier
- Tidal area
- Coastal swamp
- / Profile through tidal inlet

Profile through tidal inlet

Water levels

EHW — — — —
MHW — — — —
MSL →
MLW — — — —

EHW — — — —
MHW — — — —
MSL →
MLW — — — —

GW ∨ ∨ ∨ ∨
MSL →

EHW = Extreme high water   MLW = Mean low water   GW = Groundwater level
MHW = Mean high water   MSL = Mean sea level   ↕ Tidal amplitude

sand from the sea is transferred to the beachline. The wind then shifts the sand to the beach and dunes. If there is a surplus of sand, the beach and coastal dunes expand in a seaward direction. As well as being transported by wave action from the sea to the coastline, sand is also transported to the tidal systems by tidal and wind-driven currents. In the tidal systems, the waves and tides shape the sand banks and edges of the salt marshes. Strong winds can cause high waves to build up at the tidal inlets. The waves then erode the adjacent tidal flats at high tide and the sand that is released is deposited across the fringes of the salt marshes during storms.

Waves also shape the shores of inland lakes and tidal channels through erosion. This erosion of the shoreline was what caused the Flevo lakes and the later Zuiderzee to continue to expand in the course of the Holocene. Today, wave action plays a significant role in the erosion of the intertidal sandflat margins along the tidal channels in the Oosterschelde, where after the completion of the famous Eastern Scheldt storm surge barrier barrier the strength of the tidal currents and the supply of sediment to the tidal flats declined.

As we have observed, although the sea was the biggest contributor to the formation of the Netherlands, the influence of rivers should not be ignored.

FIG. 12 Accretion model of a tidal basin as the sea level rises. Accretion reduces the tidal volume and the maximum tide height declines, making the dried-up tidal areas suitable for habitation.

5 THE BIG RIVERS FILL THE DELTA

When we refer to the 'big rivers' in the Netherlands we mean the Rhine and the Meuse, wide water channels that make their way slowly though the landscape. But this wasn't always the case. There are different kinds of rivers. The gradient (the slope of the valley or floodplain through which a river flows), the quantity of sediment available for transport, the flow rate and seasonal changes in river discharge all determine how a river behaves and what sediments it picks up and deposits. Changes in river behaviour, especially along the Rhine and Meuse, have had a major impact on the landscape.

One kind of river is the braided river (fig. 10a). In the Netherlands these were mainly found during the last ice age, when the ground was permanently frozen ('permafrost') and there was little vegetation. Each spring the rivers discharged large volumes of water as direct run-off from melting snow. The rivers became particularly wide and shallow because the permafrost prevented the water from soaking into the soil. And because the meltwater also contained lots of sediment, the Rhine and Meuse deposited large quantities of sand and gravel in their valleys. When the water discharge was minimal, the riverbed became largely dry and more and more river channels would appear in this broad strip of sand and gravel. Sand would also be blown up from the dry riverbeds to form enormous sand dunes. We can still see them today in north Limburg and the Land van Maas en Waal. But sand dunes of this kind were also formed further westward. In the Alblasserwaard the highest tops of these dunes continue to rise above the later cover of clay and peat.

If there is relatively little sediment available and little variation in discharge, a river begins to meander, winding its way through the landscape (fig. 10b). These conditions occurred in the Netherlands from the onset of the Holocene. We can still see this phenomenon in a few streams that haven't been 'normalised' by human intervention, such as the Geul, Roer and Swalm in Limburg. These are meandering rivers that flow through valleys. They transport most of their sediment along their beds to the Meuse.

At the beginning of the Holocene the Rhine and Meuse discharged their sediment in this way to the sea, but this changed with the relative rise in sea level. Now most of the river sediment no longer found its way to the sea but was deposited before it could get there. Sand and gravel settled in the riverbed, and clay was deposited along the banks during times of flood. Thus, the floodplain rose along with the sea levels and a delta was created.

The rivers also shifted position as a result of sedimentation at the inside bends where the river water flowed quite slowly. At the outside bends, where there was an accelerated flow, material was eroded and then deposited again further downstream. This process could also lead to the appearance of small tributaries at the outside bends, which carried water and sediment to the floodplain when the river ran high (fig. 10b). If a tributary eventually took over the discharge of all the river water, an entirely new river channel would develop. This happened frequently in the Holocene, creating a delta with more than a hundred abandoned river channels. The sand in the former river channels is generally about 10 metres thick. Over time a layer of clay and peat formed outside the channels as a result of flooding. In the Alblasserwaard this layer can exceed 10 metres in thickness, but it thins out towards the east and is less than two metres at Lobith, at the German border.

In the western part of the delta, from the line linking Zaltbommel, Geldermalsen and Culemborg, the relative sea-level rise led to the emergence of a third type of river, the 'anastomosing' river (fig. 10c). This less well-known river type mainly occurs when the water level in an estuary rises, but not so rapidly as to flood the river delta completely, which would create a shallow sea or a lake. In the case of an anastomosing river, just enough river sediment is deposited to keep pace with the relative sea-level rise. This gives rise to narrow, winding rivers that barely change course, especially if there is peatland beyond the rivers. The rivers can't shift position because they are 'wedged' between metres-thick layers of clay and peat. They are unable to drain quickly at times of flood and therefore almost 'drown'. In the Netherlands, the shallow lakes that also occurred in this type of riverine landscape filled up with small organic particles and with clay and sand from the rivers. In the western Netherlands, this type of river emerged roughly between 5000 and 3000 BCE.

Thus, the relative sea-level rise had not only an impact on the position of the coastline. Rivers also changed form and location, creating layers of sand and clay in different parts of the landscape and 'filling up' the floodplain with sediment. The relative sea-level rise also set in motion – indirectly – another process that would radically change the appearance of the landscape: namely, peat formation in the river delta floodplain.

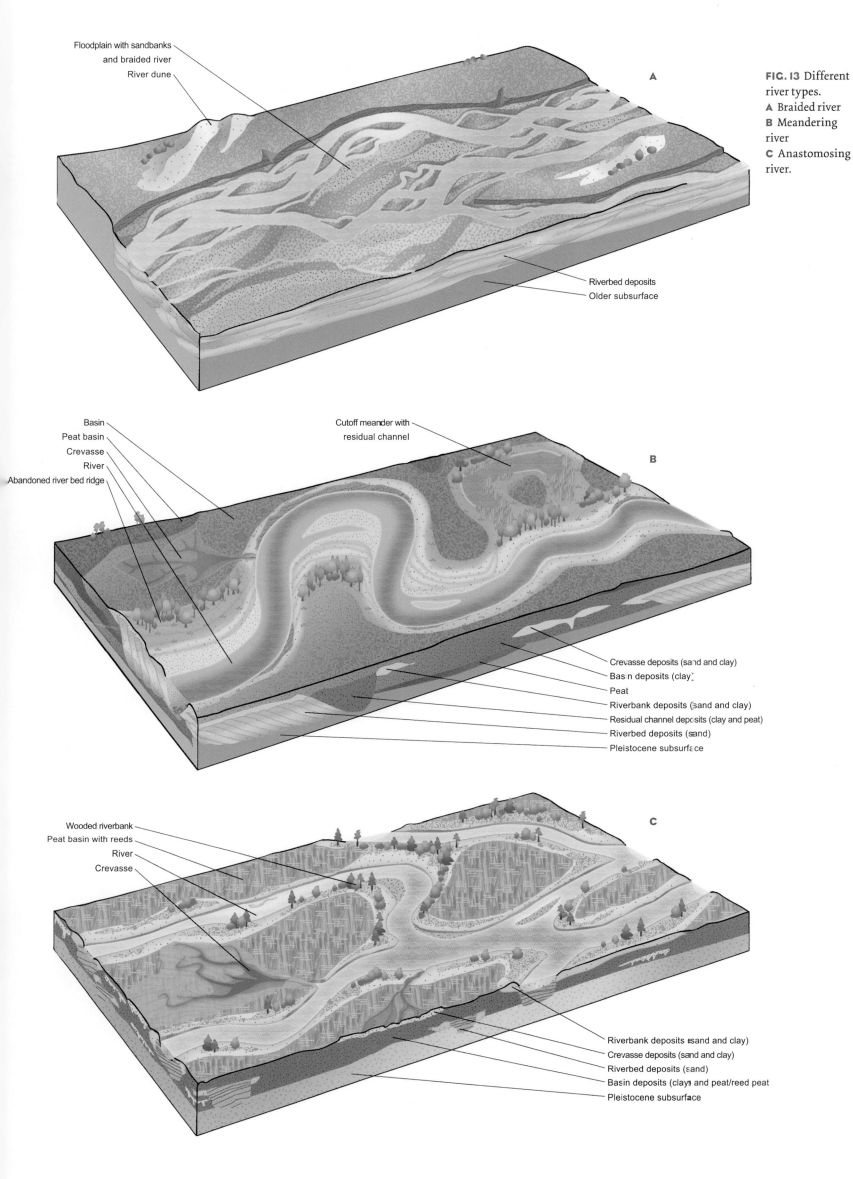

Floodplain with sandbanks
and braided river
River dune

A

Riverbed deposits
Older subsurface

Basin
Peat basin
Crevasse
River
Abandoned river bed ridge

Cutoff meander with
residual channel

B

Crevasse deposits (sand and clay)
Basin deposits (clay)
Peat
Riverbank deposits (sand and clay)
Residual channel deposits (clay and peat)
Riverbed deposits (sand)
Pleistocene subsurface

Wooded riverbank
Peat basin with reeds
River
Crevasse

C

Riverbank deposits (sand and clay)
Crevasse deposits (sand and clay)
Riverbed deposits (sand)
Basin deposits (clay and peat/reed peat
Pleistocene subsurface

FIG. 13 Different
river types.
A Braided river
B Meandering
river
C Anastomosing
river.

21

6 PEAT COVERS THE LAND

In dry conditions the remains of dead plants, shrubs and trees are normally broken down through oxidation and by worms, insects, moulds, fungi and bacteria. But when plant material ends up in a wet, oxygen-poor environment, this doesn't happen. Instead, it accumulates to form peat. The nature of the peat is determined by the original vegetation, which can range from *Sphagnum* moss or heather to shrubs and entire trees. Because the plant material is only partially decayed, it is often possible to see what the original vegetation consisted of. Aerial roots can be readily identified by their yellowish-white colour. The leaves and twigs of types of heather can also be distinguished (fig. 14). The same applies to *Sphagnum* moss, which can be recognised by the naked eye. Peat therefore provides clues about the type of landscape that existed in the area at the time when the peat formed.

Studies of peatland areas classify peat as 'eutrophic' (nutrient-rich), 'mesotrophic' (moderately rich in nutrients) and 'oligotrophic' (nutrient-poor).

Eutrophic peat is formed in places with nutrient-rich groundwater. Eutrophic peat developed in the river delta when occasional flooding brought supplies of minerals. There are different kinds of eutrophic peat. Carr peat contains the remains of alder, birch or willow from swamp forests (figs 15a and 15d). It is commonly found at some depth in the western and northern Netherlands, in land that drowned as a result of relative sea-level rise. Reed peat, another type of eutrophic peat, was formed in shallow ponds, lakes or poor-draining salt-marsh areas (fig. 15b), sometimes directly behind the beach barriers, since reed is able to survive incidental flooding by the sea. This type of peat is

FIG. 14 Peat in which leaves and twigs can still be distinguished.

also common to the western and northern Netherlands.

Mesotrophic peat is characteristic of landscapes in which the influence of the groundwater has declined and the vegetation is more dependent on rainwater, which is mineral-poor. This type of peat is represented by sedge peat. Sedge is a grass-like plant species that flourishes in very wet conditions (fig. 15c).

Finally, oligotrophic peat forms in areas where the groundwater can no longer penetrate and rainwater is the only source of moisture. When water levels are high, the peat is no longer flooded by nutrient-rich river or sea water. *Sphagnum* moss,

heather and cottongrass, species that require few nutrients, are able to grow there (fig. 15e). Today, there are almost no places in the Netherlands where oligotrophic peat formation occurs. The peatlands containing oligotrophic peat were almost all excavated in the thirteenth to the twentieth centuries. Dried *Sphagnum* peat is an excellent source of turf, and for centuries it served as a fuel, mainly for the cities.

Peat can also be eroded by the movement of water, such as currents or wave action. The collective term 'peat detritus' or 'detritus' is used to describe loose peat material in sediments.

Considering its origins, peat was mainly formed in wet

15 Landscapes featuring the different peat-forming vegetations

15A Carr

15B Reed swamp

15C Sedge swamp

15D Wet woodland

15E Raised bog

parts of the landscape, in both the high and low Netherlands. In about 7000 BCE the sea level reached the current coastline, causing the groundwater to rise in land adjacent to the sea. Seepage water from the higher sandy soils rose to the surface near the coast. A eutrophic peat layer, called basal peat, formed in this wet zone. At the bottom, at the transition from sand to peat, this peat is often made up of wood, the remains of forest that drowned as a result of sea-level rise and saturation. Basal peat itself mainly consists of reed peat, which is clayey on top and is covered by a layer of marine sediments as a result of ongoing marine drowning. The continuing sea-level rise caused the basal peat growth to shift landwards, which means the deeper basal peat is older than the basal peat that formed higher up.

Midway through the Holocene, when the sea level rose less rapidly, coastal formation switched from transgressive to regressive. In other words, the coastline no longer moved landwards but developed in a seaward direction. The tidal area behind the beach barriers became highly silted-up and began to accrete (figs 16 and 17a). Because the drainage channels in the tidal area also silted over, the salt marshes behind the beach barriers became saturated and the groundwater rose above ground level. The reed peat that formed everywhere was still occasionally flooded by the sea during severe storms. Because the peat continued to grow in height, it was inundated less and less often by nutrient-rich water. Reed-sedge peat developed as the environment became mesotrophic. In the heart of the coastal swamps, the peat reached elevations that meant it was never flooded again. From that time on it was fed only by rainwater. Because this water was poor in nutrients, the peat became oligotrophic. These nutrient-poor bogs could rise up several metres above their surroundings, in which case they are called oligotrophic peat domes or raised bogs. The entire coastal peat layer (fig. 11) that formed on the marine deposits in the western Netherlands is referred to as Holland Peat.

There was also large-scale peat formation in the delta of the major rivers during the Holocene, especially in the west where the delta was wide and the rivers lay far apart (fig. 17b). The floodplains at some distance from the rivers continued to be inundated but didn't fill up because the water contained very little sediment. The low-lying floodplains at some distance from the rivers were frequently or permanently under water, creating a clayey/peaty environment in which an alder swamp forest could evolve. Carr peat developed from the decayed wood, branches and leaves.

Lastly, peat formed in the Pleistocene sand area (fig. 17c), both in poorly drained stream valleys where carr peat developed and in higher areas that struggled with poor drainage. In kettle holes, collapsed pingos and coversand basins, the local groundwater was very high and small lakes were able to form. In environments where there was no disturbance from natural or human factors, different types of peat would often follow one another in quick succession. The vegetation and peat types could evolve from eutrophic to oligotrophic in an overgrown lake. The bottom of such lakes often had a layer of 'gyttja', a kind of muddy sediment consisting of finely distributed plant remains that had found their way into the lake and sunk to the bottom. Reed peat would then form above it until the pond was 'full'. Sedge peat represented the next stage of landscape evolution. It formed when the influence of the groundwater diminished, for example when the reed peat had completely filled up the former lake. Sometimes eutrophic wood peat would then form above this mesotrophic peat, for example when nutrients were released from the underlying peat as it dried out somewhat. Often, however, something else happened. The central parts of the lakes would become entirely dependent on rainwater for their water supply. *Sphagnum* moss would then grow, which didn't fully decompose, causing a layer of oligotrophic *Sphagnum* peat to form. Because they were not dependent on groundwater, oligotrophic peats could grow into 'peat domes', which rose up to several metres higher than their surroundings. Small peat streams carried

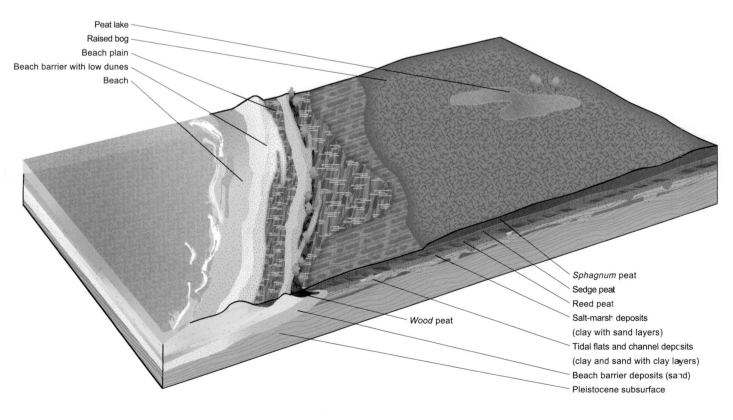

Peat lake
Raised bog
Beach plain
Beach barrier with low dunes
Beach

Wood peat

Sphagnum peat
Sedge peat
Reed peat
Salt-marsh deposits
(clay with sand layers)
Tidal flats and channel deposits
(clay and sand with clay layers)
Beach barrier deposits (sand)
Pleistocene subsurface

FIG. 16 A former tidal basin overgrown with peat.

FIG. 17 Holocene peat
formation.
A in the coastal plain
B in the river delta
C on higher Pleisto-
cene soils

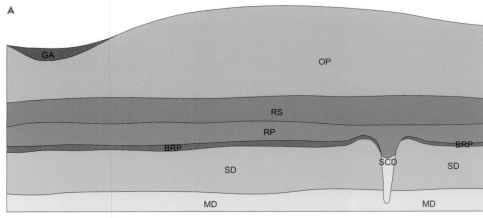

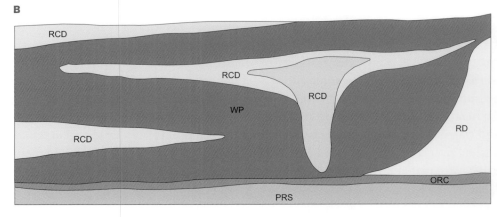

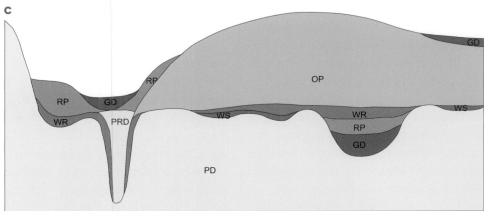

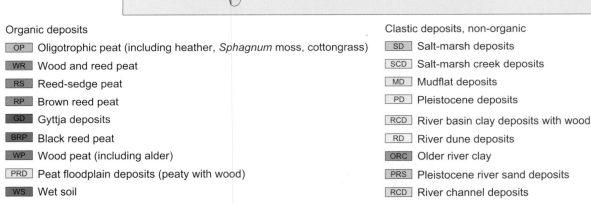

Organic deposits

OP	Oligotrophic peat (including heather, *Sphagnum* moss, cottongrass)
WR	Wood and reed peat
RS	Reed-sedge peat
RP	Brown reed peat
GD	Gyttja deposits
BRP	Black reed peat
WP	Wood peat (including alder)
PRD	Peat floodplain deposits (peaty with wood)
WS	Wet soil

Clastic deposits, non-organic

SD	Salt-marsh deposits
SCD	Salt-marsh creek deposits
MD	Mudflat deposits
PD	Pleistocene deposits
RCD	River basin clay deposits with wood
RD	River dune deposits
ORC	Older river clay
PRS	Pleistocene river sand deposits
RCD	River channel deposits

surplus rainwater to the lower streams, where the nutrients transported by the water accumulated. This then allowed eutrophic wood and reed peat to develop, or mesotrophic sedge peat.

The extensive peat formation in the Pleistocene stream valleys led to a continuing deterioration in the natural drainage on the higher sandy soils during the Holocene. As a consequence, the bogs started expanding beyond the stream valleys and low-lying sand and till areas. They would eventually cover much of the high Netherlands. During the Bronze Age they joined up with the enormous coastal peat swamps of the low Netherlands until more than half of the Netherlands was covered in peat!

After the peat formed, peat extraction and water management in peatlands have contributed to subsidence in the Netherlands (see the panel in 'Rising sea levels').

7 HUMAN INTERVENTION

Relative sea-level rise, the tide and waves, river dynamics and peat formation are the four main natural drivers behind the formation of the Dutch landscape in the Holocene. The thirteen palaeogeological maps in this atlas show how these processes have interacted with one another over time to shape the present-day landscape of the Netherlands. These maps demonstrate clearly that landscape evolution was anything but a linear process (there was, for example, both peat formation and a decomposition of peat) and that there were significant regional differences. The maps show too that humans have gradually tightened our grip on the landscape. Eventually, humans emerged to become the main driver behind the changes in the landscape of the Netherlands.

Human influence on the Dutch landscape has occurred in a number of steps. Changes in the way of life, technology, the organisation of labour and the use of natural and fossil fuels have played a decisive role. With each step the impact of humans on the natural environment has increased exponentially. Each time, a combination of effects (intended and unintended) can be identified, most of them negative. Sometimes people succeeded in overcoming these impacts through technological or organisational innovations; sometimes they didn't.

The Netherlands was inhabited during the Pleistocene, first by Neanderthals and later by modern humans. These hominids and humans lived as hunter-gatherers and had only a very limited impact on the environment and landscape.

From about 14,000 years ago, with an interruption during the Younger Dryas, flowers, shrubs and trees made their appearance (or reappearance) in the Netherlands, either independently or by taking advantage of the wind, insects or birds (fig. 19). Under human influence, the new natural Holocene vegetation rapidly underwent major changes.

What profoundly changed the relationship between people and the environment was the transition from nomadic hunting and gathering to the more sedentary hoe and plough agriculture. Deforestation and arable farming on the sandy soils had a dramatic, unintended impact on the relationship between precipitation and evaporation, causing severe depletion of the soil and rendering large areas no longer suitable for agriculture. With the introduction of the plough, it became possible to cultivate heavier soils that were less susceptible to degradation. In the space of just a few thousand years, humans had changed a closed forest landscape into an open cultural landscape, where they had left their mark virtually everywhere. For some areas in later prehistory, it has been documented that the heathlands were so intensively grazed that the vegetation disappeared altogether and sand drifts arose. This often had a detrimental effect on arable land. With the exception of these examples, however, humans were not a significant geological factor in prehistory. Their habitation options changed in response to the gradual, yet profound natural changes that are documented in this atlas. Broadly speaking, this came down to a reduction in the total living area as

large parts of the Netherlands were drowned, became waterlogged or were overlain with peat.

The Roman period is often regarded as a time when humans first intervened in the landscape and water management on a major scale. It is true that the Romans were the first who were able to organise human labour on a sufficient scale to build a coherent infrastructure of roads, dams and canals. Well-known examples are the artificial road along the imperial border, the Drusus dam and the Corbulo canal (fig. 55). Recent research shows that the Romans had to adapt their infrastructural models to local conditions in the Dutch delta, perhaps after learning the hard way.

Even more important than the activities of the Romans was the behaviour of the native population. Their numbers had grown steadily since the late Iron Age, from several tens of thousands in late prehistory to about a quarter of million in the second century CE. Increasingly, people began to inhabit and cultivate the naturally draining peatlands. However, habitation in these areas could only be sustained for one or two generations as people used artificial means to improve the drainage of the peatlands. This caused the peat to oxidise and become compacted. Homes were flooded and people were obliged to move to higher parts of the peatlands, where the process would start again. In some areas, such as Friesland, but especially Zeeland, the decomposition of the peatlands through human intervention was so extensive that the sea was able to create openings in the coast and to gain a hold on

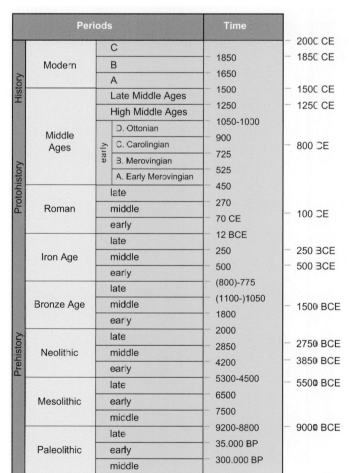

Periods			Time	
History	Modern	C		— 2000 CE
		B	1850	— 1850 CE
		A	1650	
Protohistory	Middle Ages	Late Middle Ages	1500	— 1500 CE
		High Middle Ages	1250	— 1250 CE
		early	1050-1000	
		D. Ottonian	900	
		C. Carolingian	725	— 800 CE
		B. Merovingian	525	
		A. Early Merovingian	450	
	Roman	late	270	
		middle	70 CE	— 100 CE
		early	12 BCE	
Prehistory	Iron Age	late	250	— 250 BCE
		middle	500	— 500 BCE
		early	(800)-775	
	Bronze Age	late	(1100-)1050	
		middle	1800	— 1500 BCE
		early	2000	
	Neolithic	late	2850	— 2750 BCE
		middle	4200	— 3850 BCE
		early	5300-4500	— 5500 BCE
	Mesolithic	late	6500	
		early	7500	
		middle	9200-8800	— 9000 BCE
	Paleolithic	late	35.000 BP	
		early	300.000 BP	
		middle		

FIG. 18 Geological and archaeological periods. To the right of the table are the dates of the thirteen maps in this atlas.

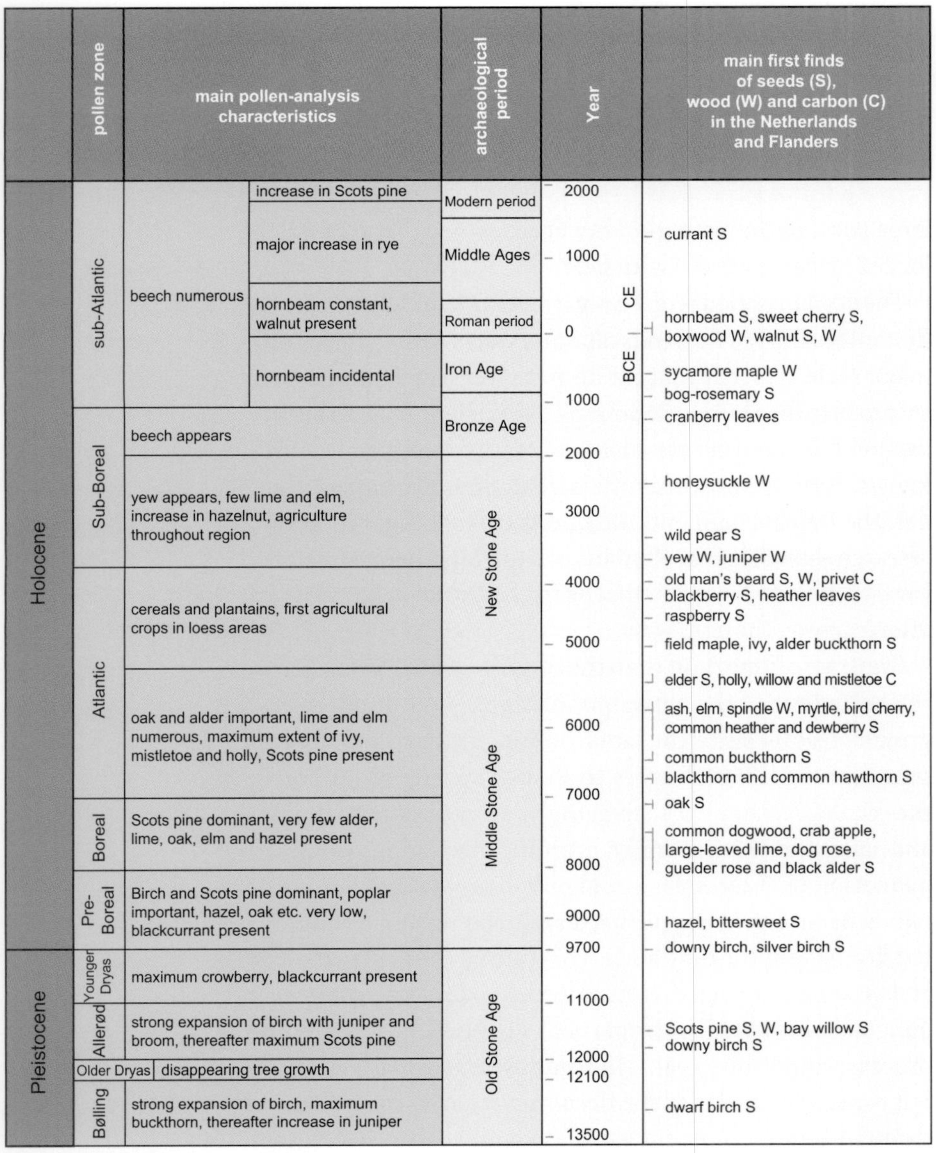

	pollen zone	main pollen-analysis characteristics		archaeological period	Year	main first finds of seeds (S), wood (W) and carbon (C) in the Netherlands and Flanders
Holocene	sub-Atlantic	beech numerous	increase in Scots pine	Modern period	2000	
			major increase in rye	Middle Ages	1000	currant S
			hornbeam constant, walnut present	Roman period	0 (CE / BCE)	hornbeam S, sweet cherry S, boxwood W, walnut S, W
			hornbeam incidental	Iron Age	1000	sycamore maple W; bog-rosemary S; cranberry leaves
	Sub-Boreal	beech appears		Bronze Age	2000	honeysuckle W
		yew appears, few lime and elm, increase in hazelnut, agriculture throughout region		New Stone Age	3000	wild pear S; yew W, juniper W
	Atlantic	cereals and plantains, first agricultural crops in loess areas		New Stone Age	4000 / 5000	old man's beard S, W, privet C; blackberry S, heather leaves; raspberry S; field maple, ivy, alder buckthorn S
		oak and alder important, lime and elm numerous, maximum extent of ivy, mistletoe and holly, Scots pine present		Middle Stone Age	6000	elder S, holly, willow and mistletoe C; ash, elm, spindle W, myrtle, bird cherry, common heather and dewberry S; common buckthorn S; blackthorn and common hawthorn S; oak S
	Boreal	Scots pine dominant, very few alder, lime, oak, elm and hazel present		Middle Stone Age	7000 / 8000	common dogwood, crab apple, large-leaved lime, dog rose, guelder rose and black alder S
	Pre-Boreal	Birch and Scots pine dominant, poplar important, hazel, oak etc. very low, blackcurrant present		Old Stone Age	9000 / 9700	hazel, bittersweet S; downy birch, silver birch S
Pleistocene	Younger Dryas	maximum crowberry, blackcurrant present		Old Stone Age	11000	
	Allerød	strong expansion of birch with juniper and broom, thereafter maximum Scots pine			12000 / 12100	Scots pine S, W, bay willow S; downy birch S
	Older Dryas	disappearing tree growth				
	Bølling	strong expansion of birch, maximum buckthorn, thereafter increase in juniper			13500	dwarf birch S

FIG. 19 The appearance of shrubs and trees in the Netherlands since about 15,500 years ago.

the hinterland. Consequently, at high tide and during storm surges large areas were lost in a very short time (compare the map of 100 CE with that of 800 CE). Unwittingly, people had caused major damage to the landscape.

After the second century we observe a gradual return to late-prehistoric conditions in demographic, economic and sociopolitical terms. This changed once again when the Dutch delta became part of the Carolingian empire. The population in the territory of the present-day Netherlands rose from several tens of thousands in the ninth century to a million by the 1500s. Underpinning this enormous demographic growth was the extensive colonisation of new areas in both the high and low Netherlands. Once again, people exploited the margins of the peatlands. This time, however, they succeeded in overcoming – through technology and organisation – the inevitable hydrological problems associated with reclamation of the peatland. They built communal quays, embankments and sluices to tackle the problems caused by oxidation and subsidence. The introduction of the windmill proved invaluable: from the early 15th century it was now possible to drain water from the polders when water levels were high. Embankments created a new problem, however. They prevented the seawater from flowing freely across the land during floods and storm surges. The decline of the storm surge storage capacity of the marshes induced a dangerous increase of the storm flood levels at the seaward side of the dikes. When a dyke was breached, the water would flood the land behind it, engulfing people and livestock and causing long-term damage. Dyke breaches are not natural disasters!

In the Middle Ages and early modern period humans harnessed their own power, together with that of animals and the wind, to become a geological factor of major significance. The saying 'God made the world but the Dutch made Holland' bears witness to this. And that statement was made even before the advent of the industrial revolution, when fossil fuels were used to rapidly and profoundly alter the landscape. The changes that the Netherlands underwent in the nineteenth and twentieth centuries would be inconceivable without the use of big ships, dredgers and excavators. Sweat was no longer the characteristic smell of Dutch engagement with the landscape, but rather the stench of exhaust fumes. The most impressive examples are of course the Afsluitdijk, the polders in the IJsselmeer and the Delta Works. However, the future of the Afsluitdijk shows the paradoxical nature of the ever-expanding opportunities we have to intervene in our environment: there are plans afoot to raise the Afsluitdijk in response to the rapid rise in relative sea level, a problem that is largely attributed to the use of fossil fuels.

8 HOW THE MAPS WERE COMPILED

As early as the sixteenth century, people in the Netherlands were aware that the southern North Sea coastal region had undergone major changes since Roman times. From their own experience and from stories passed down, they knew that the sea could engulf vast tracts of land and that silting-up could cause the coast to expand seawards. Based on landscape descriptions by classical authors such as Tacitus and Pliny, the first maps were drawn to show what our area would have looked like in the Roman period (fig. 20). It wasn't until the nineteenth and twentieth centuries that information from geological and archaeological research was included for the first time in reconstructions of this kind. This research really took off after the Second World War. In 1986 W.H. Zagwijn was first to chart the gradual evolution of the Netherlands during the Holocene in ten large-scale maps (fig. 1).

A range of disciplines have played a part in producing the palaeogeographic map reconstructions: geology, soil science, archaeology, palaeo-ecology, climate sciences, historical sciences and toponymy. The idea was to bring together and connect the knowledge gleaned from a wide range of sources. Geological research has produced data from hundreds of thousands of soil core samples that show how the subsurface is built up of different sediments. They are indicators of the landscapes in which these sediments were deposited: sea, mudflats, salt marshes, beach, dune, peat, river, floodplain or sand drift. Today, the LIDAR-based surface elevation data in the *Actueel Hoogtebestand Nederland* (AHN) is a useful tool for identifying old landscape forms that still lie on the surface. Most important of course is the dating of sediments, for which geologists have various scientific methods at their disposal. Archaeology is also important in this respect. Studies of archaeological sites often give us a clear picture of changes in the landscape, flora, fauna and habitation possibilities – all accurately pinpointed in time. Historical land maps and sea charts are an important source for the period from the sixteenth century onwards.

However, map compilation involves more than simply combining information from different sources. Information only acquires meaning if experts can place it in a model showing the relationship between the key factors that drive coastal and landscape evolution. Such a model is also important for areas for which we have little data. For example, we know that sea-level rise has been the key driving factor in the Holocene Netherlands. Especially for older periods, for which relatively

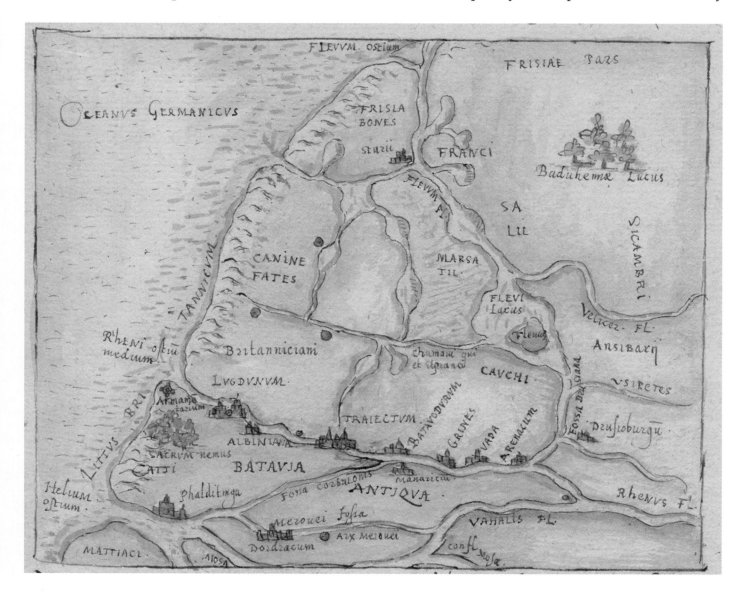

FIG. 20 Manuscript map of the Netherlands in the Roman period by Arnoldus Buchelius (1565-1641).

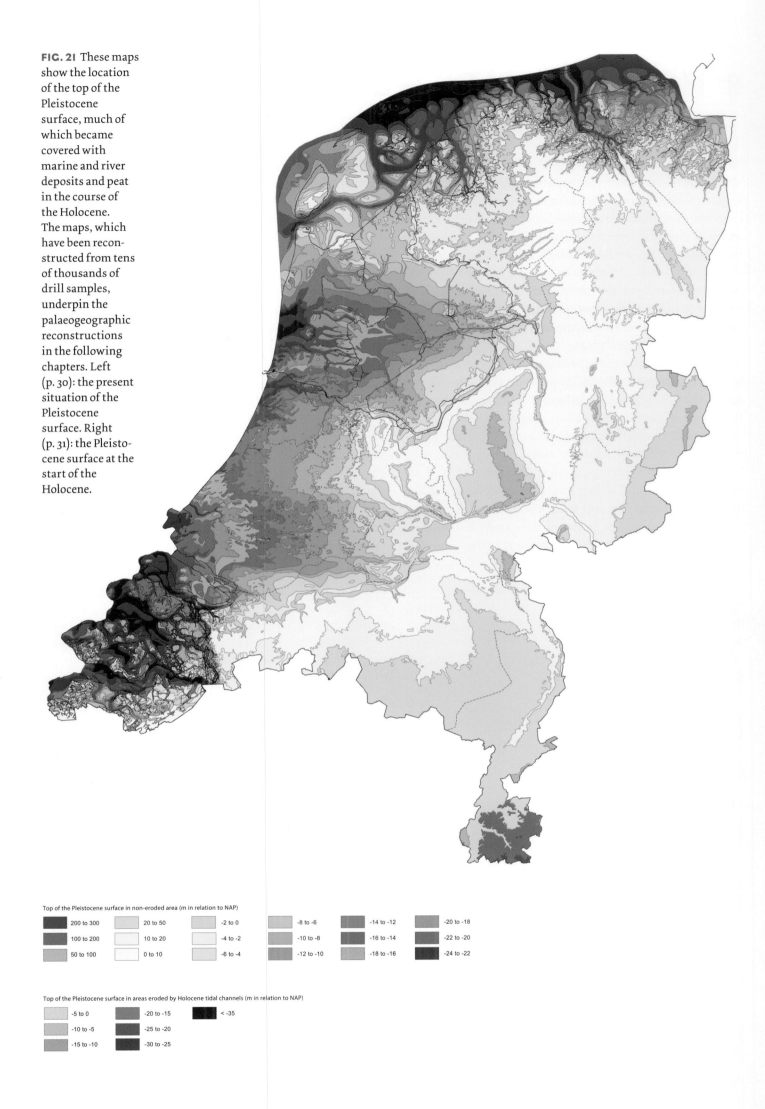

FIG. 21 These maps show the location of the top of the Pleistocene surface, much of which became covered with marine and river deposits and peat in the course of the Holocene. The maps, which have been reconstructed from tens of thousands of drill samples, underpin the palaeogeographic reconstructions in the following chapters. Left (p. 30): the present situation of the Pleistocene surface. Right (p. 31): the Pleistocene surface at the start of the Holocene.

Top of the Pleistocene surface in non-eroded area (m in relation to NAP)

200 to 300	20 to 50	-2 to 0	-8 to -6	-14 to -12	-20 to -18
100 to 200	10 to 20	-4 to -2	-10 to -8	-16 to -14	-22 to -20
50 to 100	0 to 10	-6 to -4	-12 to -10	-18 to -16	-24 to -22

Top of the Pleistocene surface in areas eroded by Holocene tidal channels (m in relation to NAP)

-5 to 0	-20 to -15	< -35
-10 to -5	-25 to -20	
-15 to -10	-30 to -25	

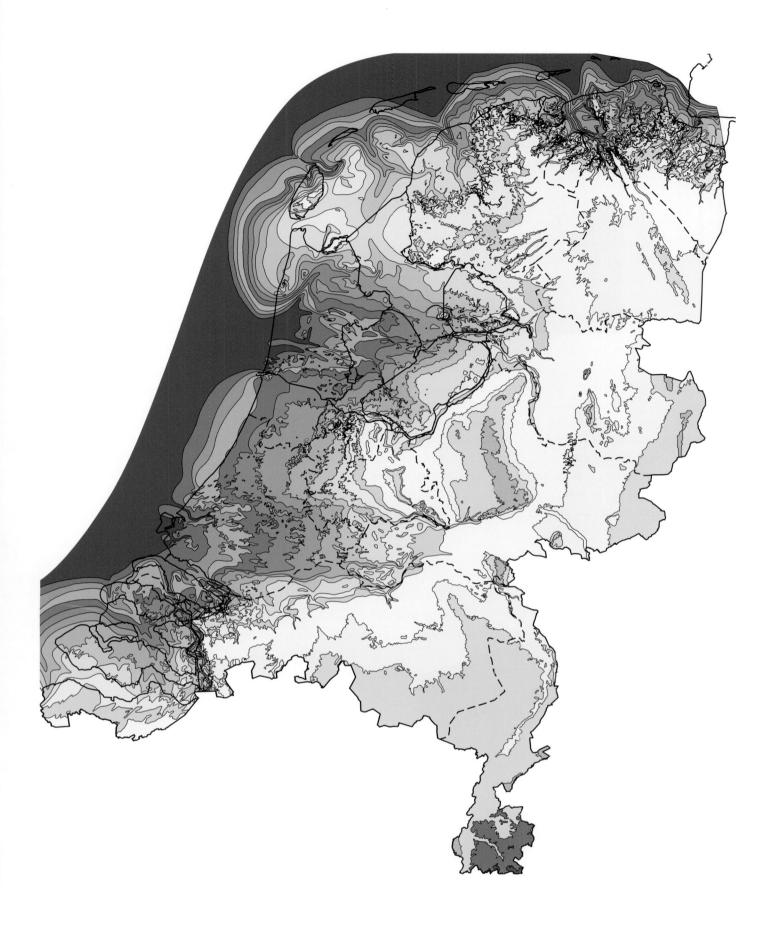

little data is available, the map image of the western and northern Netherlands has largely been arrived at by combining data from the elevation map of the Pleistocene surface at the start of the Holocene with data about sea-level rise. Thus, the gradual drowning of the western and northern Netherlands can be illustrated in spatial terms. Another example concerns the development of peat areas. Because of peat excavation, the position and nature of these peat areas is known to us only partially – if at all – from historical maps and sources. Thanks to our knowledge of the hydrological conditions in which different types of peat develop, we can once again restore this vanished peatland to its rightful place on the map.

Palaeogeographical reconstructions can be made on different scales. New and detailed geological and archaeological research is often carried out in the context of road and rail construction, new housing projects, the development of industrial areas, nature development or sand removal. This provides the basis for local reconstructions of landscape evolution. Local reconstructions can be included in regional mapping (fig. 23). Regional maps of this kind are now avail-

31

FIG. 22 The area around Castricum (approx. 2.5 x 2.5 km), where the Oer-IJ discharged into the North Sea in the Iron Age. The map on the left shows the LiDAR surface elevation data of the Netherlands (AHN), an important tool in palaeogeographic reconstructions, for this area. This highly detailed elevation data can help determine the boundaries of the different landscape units. The map on the right shows the regional palaeogeographic reconstruction of the same area in about 100 CE. The beach barriers, beach plains and North Sea beach are shown in yellow; the various landforms on the salt marshes are shown in green; and the brown colours represent the peat areas.

Height in metres (AHN)

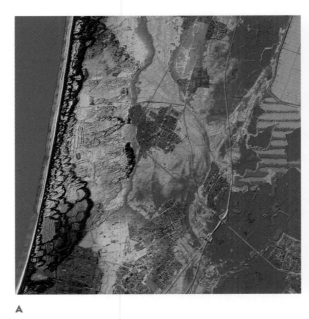

A

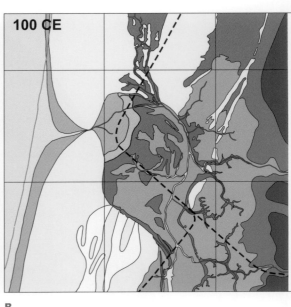

B

able for large parts of the Netherlands, such as Zeeland, the river region, the Oer-IJ area (fig. 22), the former Almere region and large parts of Friesland and Groningen. These regional reconstructions can then be merged into a national one, as was done in this atlas. In the case of areas for which less informa-

tion is available, the maps are supplemented by expert knowledge. Therefore, they should never be enlarged for use at a regional or local level. A more detailed explanation of the map compilation for this atlas can be found in Peter C. Vos, 2015: *Origin of the Dutch Coastal Landscape*, Groningen.

FIG. 23 Local palaeogeographic reconstructions for the 'Vergulde Hand-West' archaeological excavation site at Vlaardingen (approx. 12 ha). In around 250 BCE this was an inhabited peat landscape with narrow creeks. In 100 CE the area had become inundated and it changed into a less densely inhabited landscape of floodplains (ochre) and mudflats (light green) with tidal channels and creeks. The remains of farmsteads are marked with a symbol. The grid shows the parcelling pattern in the Roman period. The excavated areas are outlined in black and numbered.

32

9 NOTES ON THE MAP LEGENDS

PLEISTOCENE LANDSCAPES

floodplain and stream valleys At the start of the Holocene, the Rhine and Meuse floodplain and the stream valleys contained sand and gravel that was deposited by braided rivers and streams. The rivers and streams began to meander at that time and little new sediment was deposited. However, the old sediment was constantly reworked as the meanders shifted.

16 m below NAP	0 – 16 m below NAP	above NAP

coversand area The surface of the coversand area largely consists of coversand that was deposited by the wind. In the northern Netherlands this is often a thin layer overlying the till. The till dates from the second-to-last ice age, when this part of the Netherlands became covered with land ice. The coversand was deposited during the coldest part of the last ice age, when the land ice didn't extend as far as the Netherlands, but the country was part of a polar desert with sparse vegetation. This legend item classifies areas according to their elevation in relation to NAP (Amsterdam Ordnance Datum). The lowest-lying areas were the first to be flooded in the Holocene. In addition to coversand, this legend item also includes other Pleistocene deposits, such as loam and small peat bogs. From 1500 CE parts of the coversand turned to drift sand.

river dunes The river dunes were also formed in the last ice age. They consist of sand, frequently coarse, that has blown from the dry beds of braided rivers. This happened when water levels were low, such as in summer. Small dunes of this type are also found along most stream valleys; they are not marked on the maps but are shown as coversand areas).

loess area In these regions the loess lies on the surface. Like coversand, loess was deposited by the wind during the last ice age. However, loess is much finer – and more fertile – than coversand or river-dune sand.

drift-sand areas In the (late) Middle Ages and premodern times heath lands were used intensively for sod cutting and grazing. They became overexploited. The loss of vegetation made way for the wind to produce large drift-sand areas.

ice-pushed ridges, ice-pushed till and ridges and valleys shaped by flowing land ice These elevations in the landscape were formed during the second-to-last ice age, about 150,000 years ago, when the Netherlands was partly covered in land ice. The ice tongues pushed up older river deposits from the subsurface to form hills (ice-pushed ridges). Till, created when sediment is finely ground under the weight of the ice cap, formed under the ice. The till also contains grains of sand, gravel, pebbles and stones transported from Scandinavia by the land ice. Parts of the till are also pushed up and form part of the glacial hills and ridges. The higher ice-pushed ridges of Gaasterland are partly comprised of till. Glacial ridges and valleys were also formed as a result of the scouring effect of the flowing land ice. The parallel valleys and ridges in Drenthe, including the Hondsrug, are an example of this. In some instances, the features in this legend item are covered by a thin layer of younger coversand.

areas with Tertiary and older deposits The subsurface in these areas consists mainly of clay and limestone that was deposited before the start of the Pleistocene, 2.6 million years ago. These areas are found in Twente, the Achterhoek and southern Limburg, where the old deposits lie on the surface. This is because these areas rose slowly in the course of the Pleistocene, causing the original younger deposits to wear away through erosion.

HOLOCENE LANDSCAPES

beach barriers and low dunes Before CE 900 there were no high dunes along the coast, but instead beach barriers and low dunes up to 6 m +NAP. These were formed when wave action carried sand from the seafloor to the coast.

high dunes High dunes were formed after CE 900. Sometime before, a change had occurred directly off the North Sea coast. The gently sloping seabed had probably become much steeper as a result of erosion. The sand that was worked loose was then swept by wave action onto the coast in much greater quantities than before. From that time on, large sand drifts were formed into high parabolic dunes by the wind. Humans were probably also a major cause of these drifting sands because they had profoundly disturbed the dune vegetation.

beach plains and dune valleys Beach plains are located between the beach barriers. Low-lying former beaches are covered for the most part in a thin layer of peat, which is not shown on the maps.

sand- and mudflats Intertidal area that is flooded at high tide and dry at low tide.

salt marshes and floodplains Tidal areas and floodplains are directly influenced by the sea and rivers respectively. Salt marshes are located above the mean high water level and are only inundated during spring tides and storm surges. Flood plains are flooded when a river bursts its banks due to high water flows. Mainly sand is deposited in the immediate vicinity of the river channel, while predominantly clay is deposited further from the river (and also peat). In view of the national scale of the reconstructions, salt marshes and floodplains have been combined.

areas of salt-marsh ridges and levees Parts of the salt marsh that rise only a few decimetres above their surroundings. These relative elevations form along the edges of salt-marshes and creeks as salt-marsh ridges and tidal levees. The relative difference in height between a channel and salt marsh can sometimes be reversed later if the sandy fill of creeks compacts to a lesser degree than the surrounding salt-marsh clay and peat deposits. These 'inversion ridges' are also included here.

peat areas The soil in the peatlands is primarily composed of peat. Because large quantities of peat were excavated in past centuries for use as fuel, it is very difficult to indicate the exact expansion of the peat area in the periods before 1500 CE. This means that what is marked as peat on the map may have partly been made up of wet stream valleys and wet heath; peat might not actually have formed there. The peatland on the map of 2000 CE includes areas of residual peat – in other words, places where peat was excavated after 1850, but where until recently the ground still contained residual peat. In the second half of the twentieth century, the groundwater table was lowered through large-scale land consolidation. This then sometimes triggered oxidation, as a result of which much of the residual peat has since disappeared.

embanked salt marshes and river plains In the eleventh century, people started building dykes and embankments on a large scale along the coast and in the river region. These embanked areas appear as embanked tidal areas and floodplains on the maps of 1250, 1500, 1850 and 2000. For the coastal region, a distinction can be seen between the high- and low-lying areas.

reclaimed lake Large lakes and ponds began to appear in the peat areas from the Middle Ages onward as a result of peat dredging and natural peat reduction in lakes. The advent of windmills made it possible to drain these lakes, giving rise to the 'reclaimed lakes' we see on the maps of 1850 and 2000.

towns and cities Towns began to appear from 100 CE onwards. The map of 100 CE shows two Roman towns – Noviomagus (Nijmegen) and Forum Hadriani (Voorburg) – although Forum Hadriani didn't acquire formal city rights until somewhat later. On the map of 800 CE we can see the main centres of trade at that time. These towns hadn't yet been granted city rights. Cities with city rights are shown on the maps of 1250, 1500 and 1850.

urban area With the exception of Amsterdam, cities were so small before 1850 that they would be invisible on the map. They are therefore marked with a symbol on the maps up to 1850. It is not until the maps of 1850 and 2000 that the built-up area had become large enough to be marked as 'urban area'.

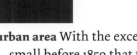

outer water and inner water The outer water comprises brakish water and salt water from (sub)tidal environments (North Sea, tidal inlets and channels, and lagoons). All the remaining water is termed inner water. It encompasses rivers (including freshwater tidal zones), lakes, ponds, embanked sea inlets and canals.

THE MAPS

9000 BCE
RISING TEMPERATURES

FIG. 24 The Kobuk River, Alaska. The landscape of the major rivers may have looked like this in the late Pleistocene.

The map of 9000 BCE shows the Netherlands shortly after the last ice age, the end of which marked the transition from the Pleistocene to the Holocene. The temperature rose rapidly at that time and yet the sea level was still several dozen metres below what is today. This is because huge volumes of water were still trapped in the melting ice caps. The North Sea was still largely dry land. This area hasn't been included in the reconstruction because there is much less data available for it. What is the now the bottom of the North Sea was at that time home to fishers, hunters and gatherers. Small bands of people also roamed across the territory of the present-day Netherlands in search of food.

Figure 5 in the Introduction clearly shows that the sea level was lower than it is today, with the coastline much further to the west than today's coastline. Much of the present-day North Sea was dry land and the British Isles weren't yet an island but were attached to the European mainland.

And yet the differences from today's sea level were not equally big everywhere. In the Strait of Dover the sea level was 26 metres lower than it is now, while at the Doggersbank further to the north it was 55 metres lower. These differences are the result of geological processes that took place during the past 11,000 years. The central part of what is now the North Sea experienced a more rapid subsidence than in the south because of the melting Scandinavian ice cap. As the weight of the ice diminished, the land began to rise, a process that is still continuing. At the same time, the Earth's crust around Scandinavia sank (see figs 6 and 9).

The soil at that time was also substantially different from today. At the start of the Holocene, wind and water began to cover the Pleistocene land with all manner of younger deposits, especially sand and gravel. To find out what the soil looked

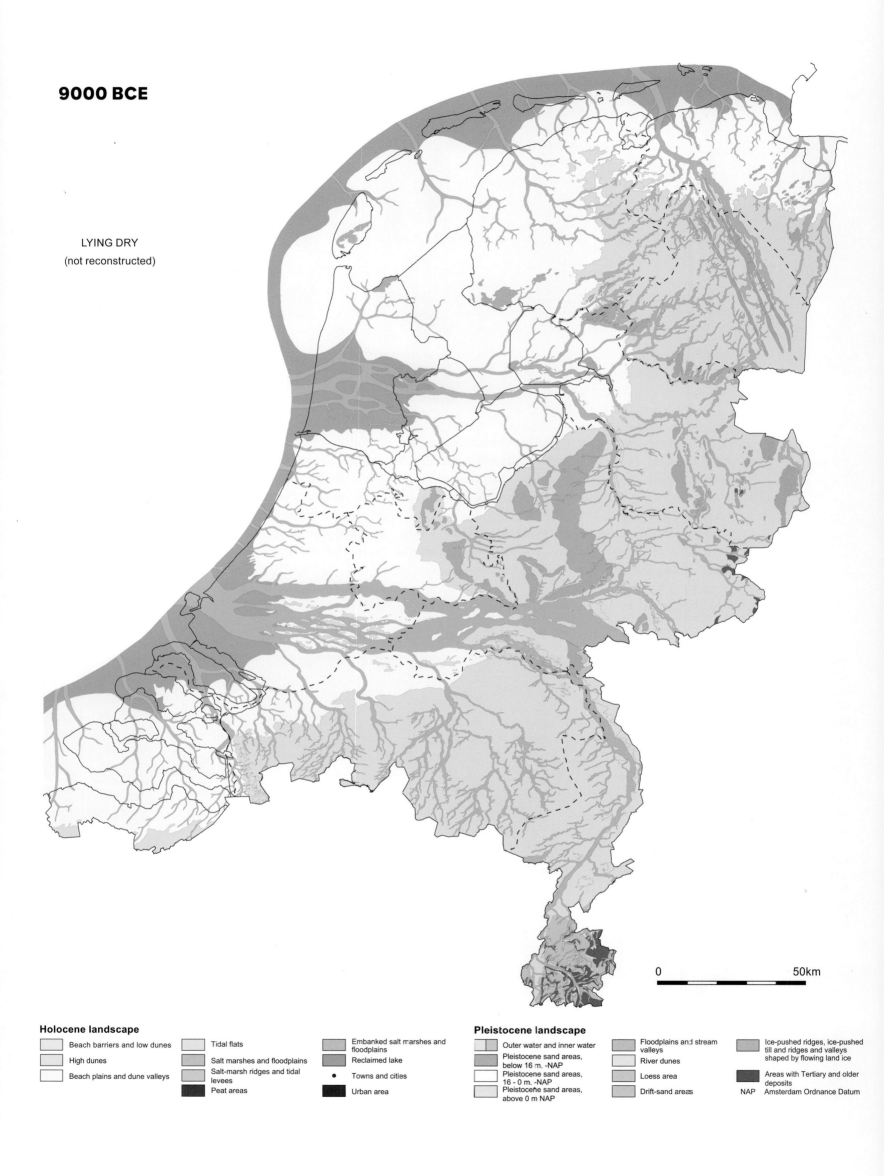

9000 BCE

LYING DRY

(not reconstructed)

0   50km

Holocene landscape

Beach barriers and low dunes

High dunes

Beach plains and dune valleys

Tidal flats

Salt marshes and floodplains

Salt-marsh ridges and tidal levees

Peat areas

Embanked salt marshes and floodplains

Reclaimed lake

• Towns and cities

Urban area

Pleistocene landscape

Outer water and inner water

Pleistocene sand areas, below 16 m. -NAP

Pleistocene sand areas, 16 - 0 m. -NAP

Pleistocene sand areas, above 0 m NAP

Floodplains and stream valleys

River dunes

Loess area

Drift-sand areas

Ice-pushed ridges, ice-pushed till and ridges and valleys shaped by flowing land ice

Areas with Tertiary and older deposits

NAP   Amsterdam Ordnance Datum

FIG. 25 Arrows with flint arrowheads the size of a fingernail were part of a hunter's equipment during the Mesolithic (c. 9000-5000 BCE). These 'microliths' are the most common artefacts discovered at sites from this period.

like 11,000 years ago, we have to 'peel back' these younger Holocene deposits from the present-day surface. This map shows the resulting picture. Subsequent maps in this atlas have been made by replacing the Holocene deposits step by step over this 'foundation' map.

In the low-lying parts of the Netherlands almost all the landscape elements we see here were overlain with sediments during subsequent millennia. This process happened first in the most low-lying areas.

During the last ice age the Rhine and Meuse were broad, braided rivers that discharged highly variable quantities of water. As a result of climate warming, they became meandering rivers with a more stable discharge.

As well as the climate, the natural environment also changed radically. Until 9700 BCE the area had been characterised by an open tundra vegetation, but this landscape gradually gave way to forests.

Initially still cold, the Netherlands at that time was the habitat of nomadic hunters, fishers and gatherers. We make a dis-

FIG. 26 Reconstruction of a screen belonging to Mesolithic hunter-gatherers (Archeon, Alphen aan den Rijn). In Geldrop, North Brabant, indications were found of a similar screen made of preserved skins coloured with ochre. This screen belonged to the late-glacial Ahrensburg culture.

tinction here between people from before 9700 BCE, the period that archaeologists label the Upper Palaeolithic (late phase of the early Stone Age), and the inhabitants of after 97000 BCE (Early Mesolithic or early phase of the Middle Stone Age).

The last people from the *Upper Palaeolithic* belonged to what is known as the Ahrensburg culture. Their traces are mainly found in the sandy region of Brabant and Limburg. We know from other countries that reindeer were the chief food source for these hunters, who also hunted horses and 'cold' species such as arctic foxes and hares. When hunting, they made use of dogs, the oldest domestic animal. It is conspicuous that most finds have been made on high ridges along former lakes and watercourses, where the food supply was probably larger and more diverse than elsewhere in the landscape. This would have been a reason for prehistoric hunters to return there regularly. The less varied parts of the landscape, such as the vast coversand plateaus or plains, have yielded almost no finds at all.

We find habitation traces from the *Early Mesolithic* not only in the southern Netherlands, but also on the higher sandy soils in the north. The perishable huts or tents of these nomads have never been found, but their stone and flint tools have survived. Splinters and fragments point to the working of stone and flint, while burnt stones indicate the presence of camps. At some locations, we find pits that were used as fireplaces. Flecks of red pigment (ochre) have also been found, although we are unable to establish with certainty what they signify. One very special find is that of a canoe at Pesse in Drenthe. It has been preserved thanks to its location in a peat-filled stream bed rather than in sandy soil.

When reindeer disappeared from the landscape, hunters turned their attention to other quarry, such as elk, red deer, aurochs and wild pigs. Unfortunately, the remains of these hunting spoils have completely decayed in the higher sandy soils of the Pleistocene Netherlands. Only in a silted-up meander at Zutphen have the remains been excavated of red deer, wild boar, roe deer, bears, wild cats and beavers, together with several tool fragments from the Early Mesolithic. Hazelnuts are the only plant food remains that we know of from that time. The oldest date to about 8300 BCE.

No find sites from this period are known from the western Netherlands and the present-day North Sea area. And yet we know that hunters, gatherers and fishers were active there too: many bone harpoons have been found in the sand dredged up from the North Sea bed and brought to the Maasvlakte in Rotterdam. These harpoons show that people once hunted west of the current coastline, before the area drowned and became the North Sea. It is no coincidence that there have been so few other finds in the western Netherlands, where the landscape was overlain in the course of time with thick layers of peat and clay. This makes it almost impossible to find traces of habitation, much more so than on the higher Pleistocene sandy soils, where the find layers lie close to the surface.

We know very little about regional differences in habitation: the number of excavated find sites is simply too small. At present we are also completely in the dark about the size and composition of these communities. Nor do we know anything about how they dealt with their dead. No burials from the Upper Palaeolithic or Early Mesolithic are known in the Netherlands.

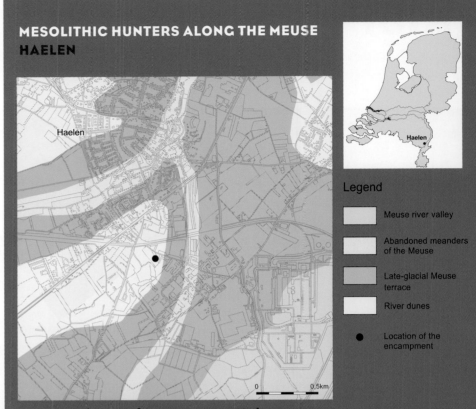

MESOLITHIC HUNTERS ALONG THE MEUSE
HAELEN

Legend

Meuse river valley

Abandoned meanders of the Meuse

Late-glacial Meuse terrace

River dunes

● Location of the encampment

FIG. 27 Location map of an encampment at Haelen (8300-8200 BCE).

From 8300 to 8200 BCE there was a small encampment at the village of Haelen in central Limburg. Probably comprising no more than one or two huts or tents, it occupied a sandy ridge not far from a meander of the Meuse that nowadays is silted up.

The small band of hunter-gatherers who used this camp probably chose the site for the wide range of food sources it offered. They most likely hunted animals that lived in the open forest, such as aurochs, red deer and wild boars. They also caught fish in the Meuse and ate plant foods, including roasted hazelnuts.

Another advantage of this location was its proximity to river deposits containing lots of gravel. They used the flint that they collected there to make stone tools, such as arrowheads. Because these tools are so small (no more than 2 cm), we refer to them as 'microliths'.

'Haelen' wasn't the only Mesolithic camp in this region. Camps from that period have also been found at other places along the Meuse.

5500 BCE
RISING WATER LEVELS

FIG. 28 Dense deciduous forest in Białowieski National Park in Poland. Whereas large parts of the low Netherlands drowned, much of the high Netherlands was covered in this type of forest during this time.

From 9000 to 5500 BCE the sea level rose very rapidly. In about 7000 BCE the British Isles became an island and 1500 years later the area we now call the Netherlands came to lie on the coast. The groundwater table also rose, creating vast tracts of peatland in the west of the Netherlands from 7000 BCE onward. Hunter-gatherers made efficient use of the new opportunities offered by the richer landscape.

The fact that we have compiled a single map to cover such a long period (from 9000 to 5500 BCE) doesn't mean that nothing happened during that time. We simply don't have enough information at our disposal to create a reliable map image for the intervening period.

What we do know for certain is that the sea level rose very quickly, by no less than 60 to 75 cm per century. This also led to a rise in the groundwater table. In low-lying areas this process was further intensified by the supply of seepage water from the higher sandy soils, causing the groundwater to reach ground level in many places. Bogs developed and peat began to form as plant remains were submerged and could no longer decompose.

In about 7000 BCE the first coastal peat bogs appeared in the lowest parts of the big river valleys, at about 20-25 m below NAP (still above sea level at that time). As the sea level continued to rise, the coastal peat bog shifted further landwards. Over time the sea flooded the coastal peat with increasing frequency, leaving behind marine clay on top of the peat. This process also began in the low parts of the river valleys of the

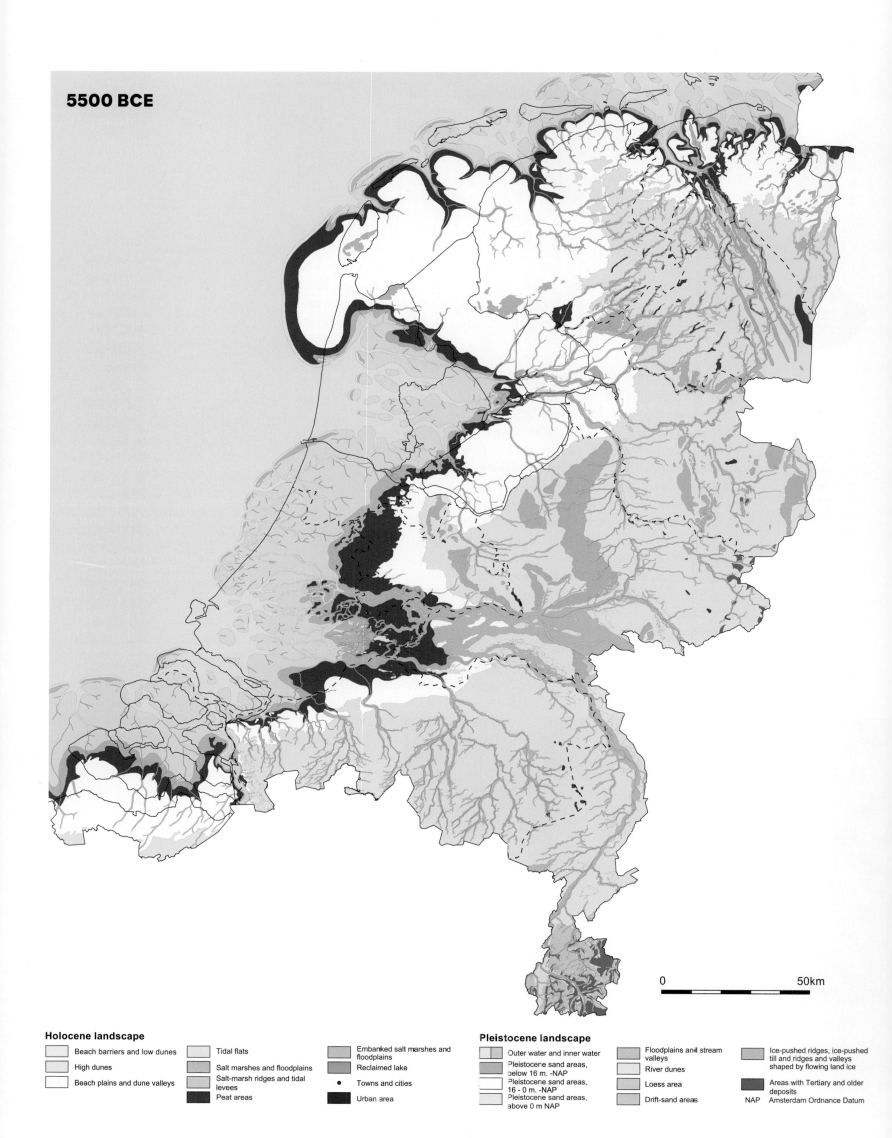

5500 BCE

0   50km

Holocene landscape

- Beach barriers and low dunes
- High dunes
- Beach plains and dune valleys
- Tidal flats
- Salt marshes and floodplains
- Salt-marsh ridges and tidal levees
- Peat areas
- Embanked salt marshes and floodplains
- Reclaimed lake
- Towns and cities
- Urban area

Pleistocene landscape

- Outer water and inner water
- Pleistocene sand areas, below 16 m. -NAP
- Pleistocene sand areas, 16 - 0 m. -NAP
- Pleistocene sand areas, above 0 m NAP
- Floodplains and stream valleys
- River dunes
- Loess area
- Drift-sand areas
- Ice-pushed ridges, ice-pushed till and ridges and valleys shaped by flowing land ice
- Areas with Tertiary and older deposits
- NAP   Amsterdam Ordnance Datum

FIG. 29 In the swampy Rhine and Meuse delta, the Mesolithic hunting and fishing population was dependent on transport by water. Canoes made from hollowed-out tree trunks appear to have been the standard vessel. This 5.5-metre-long example, made from the trunk of a linden tree, was excavated at Hardinxveld-Giessendam. It dates from about 5000 BCE.

Rhine and Meuse, Scheldt and Oer-Vecht, although there were significant regional differences.

Because the Rhine and Meuse were large rivers that carried lots of material, they deposited sand and clay along the banks in fairly large quantities. The Rhine and Meuse estuary was raised higher and higher and peat formed between the river channels. This meant that sea flooding was now confined to the area west of Dordrecht-Rotterdam-Gouda.

This process was quite different from what happened in the Oer-Vecht valley (which at that time discharged into the sea at Alkmaar) and the Scheldt valley. Because there were no large rivers there, much less sediment was available and sediment deposition couldn't keep pace with the rising sea level. As a result, all parts of these valleys that were lower than 8 m below NAP became submerged between 7250 and 6500 BCE. Because there was little tidal movement at the coast (1 m or less), a brackish lagoon environment developed. The calm nature of this environment meant that the clay built up in thin layers.

FIG. 30 Reconstruction of a Mesolithic hut (Archeon, Alphen aan den Rijn). Only a small number of Mesolithic hut floorplans have been excavated in northwestern Europe. In most cases, these are circular to oval huts measuring 5 to 7 m in diameter. They provided shelter for a single family.

This is harder to establish with certainty in the case of the Scheldt valley, where later erosion removed much of the clay.

The situation was different again in the flooded valleys of the Hunze and Fivel, where the clay layers (16 to 20 m below NAP today) can be several metres thick. This tells us that the deposition of clay kept pace with the sea-level rise. Here too the clay has a different composition: it is much less layered, but it is humus clay, which means that it contains partially decayed plant remains. From the presence of plant roots we can deduce that this landscape was dry at regular intervals and therefore covered in vegetation.

By around 5500 BCE the sea level had risen to 6 to 8 m below NAP. Several kilometres beyond the present-day coastline were beach barriers. The landscape behind the barriers was more or less comparable to today's Wadden area. The most low-lying Pleistocene valleys and sandy regions of the western and northern Netherlands had become large tidal basins. The sea level at that time was still rising at about 40 to 50 cm per century. The rise was smallest in Zeeland and biggest in the northern Netherlands, where the ground was subsiding more quickly. In the low-lying central parts of the basins, sedimentation couldn't keep up with the sea-level rise, which meant that – unlike today's Waddenzee – they were permanently under water. The higher margins of the basins evolved into a tidal landscape with tidal channels, mudflats and salt marshes. The neighbouring peat areas – formed when the rising sea level pushed up the ground water – rose to a height of about 5 to 6 m below NAP.

As a consequence of the still rapidly rising sea level, the Rhine and Meuse river delta shifted eastwards during this period. This created new river channels, which diverged frequently and reconnected again elsewhere. In the flood plains between these channels, swamp forest appeared, where (clayey) wood peats could develop. The wood peat area, which covered much of the total river delta area, consisted mainly of alders.

The river delta contained Pleistocene river dunes, which were high, dry locations. Because the dunes were completely covered in vegetation, they weren't eroded by the encroaching water. However, because of the rising water table they were increasingly covered with peat and river clay over time. It wasn't long before the smaller, low dunes were covered entirely, leaving the higher dunes protruding like dry 'islands' from the marshy delta. These islands are referred to in Dutch as *donken*.

Further upstream in the Rhine and Meuse river valleys, the deposition of clay and sand continued. In the eastern part of the Betuwe (the Over-Betuwe) the river channels still occupied their old valleys.

Peat formation occurred not only in the coastal zone of the low Netherlands and the Rhine and Meuse river delta. Peat also formed locally in the high Netherlands, in areas with poor drainage or where seepage occurred. Peat development in these areas wasn't directly influenced by the rising sea level. Areas where groundwater stagnation occurred and where large bog complexes developed during the course of the Holocene include the Peel (Brabant), southeast Drenthe and southeast Groningen.

Thus in the period from 9000 to 5500 BCE some areas saw far-reaching changes, which were mainly linked to the increasingly wet conditions. But outside those areas the physical Dutch landscape displayed a remarkable continuity, with very little actual change in much of the high Netherlands for a period of some 3500 years. What did change markedly, however, was the landscape vegetation and the opportunities this offered for human food supply. From 9000 to 5500 BCE the landscape evolved from open steppes with drift sands, to pine and birch forest, and then to mixed deciduous forest with oak and linden.

The filling-up of the North Sea basin brought significant changes to the landscape. Large habitable areas disappeared under water, while other coastal areas became much wetter.

Despite this, there were few changes in the lifestyle of the people in this part of Europe. They remained nomads who lived in temporary, seasonal camps and whose livelihood centred around hunting, fishing and foraging for roots, nuts and fruit. They could do so because the population density had been low for a long time and their territories were very large (probably many hundreds, if not thousands of square kilometres). This meant they could cope with the 'pressure on space' without too much difficulty. The development of extensive marsh landscapes also led to a greater landscape variation and to greater diversity of flora and fauna, which increased their opportunities for subsistence.

In this period too, the Netherlands wasn't inhabited with equal density everywhere. In the oldest period (9000-7000 BCE) certain activities appear to have been concentrated on the river dunes, on coversand hills along the larger stream valleys and in what are now the peat colonies in Groningen and Drenthe. The emphasis slowly shifted to areas along the eastern Overijsselse Vecht (8000-5500 BCE) and the Oer-Vecht region of Flevoland (7000-5000 BCE). We also observe such shifts in the southern Netherlands, including in the Meuse valley. There seems to be a link to landscape changes and therefore to opportunities for exploitation.

The fauna we know primarily from two late Mesolithic hunting camps on river dunes in the then Rhine-Meuse delta in the Alblasserwaard. The main prey animals found there were wild boar, red deer and fur animals (otters and beavers). People also occasionally hunted pine martens, polecats, wild cats, aurochs, roe deer and elks, supplemented by a broad spectrum of birds, fish and the occasional terrapin. Dating from the same period is the butchering camp at Jardinga (in the municipality of Ooststellingwerf in Friesland), where aurochs and red deer were slaughtered. The hunter-gatherers also lived on plant food sources such as hazelnuts, berries from the blackberry family and water chestnuts (aquatic plants that now only grow in southern Europe).

Hunter-gatherers made extensive use of natural resources. This doesn't mean, however, that they left their environment untouched. They had quite some impact on the landscape at a local level, for example by felling pine and birch trees for the production of wood tar in specially dug 'fire pits'. Wood tar was used as a 'glue' in the manufacture of tools. There is also growing evidence for the deliberate burning-off of tracts of land, which resulted in a more open vegetation and different flora and fauna. People therefore had both a direct and indirect impact on the landscape, albeit only at a local and regional level.

Although the landscape underwent major changes at this time, some locations were in use for remarkably long periods, often for many generations; these are referred to as 'persistent places'. However, the activities carried out there weren't necessarily always the same. A location may have been used for the production of wood tar on one occasion and as the site of a hunting camp on another.

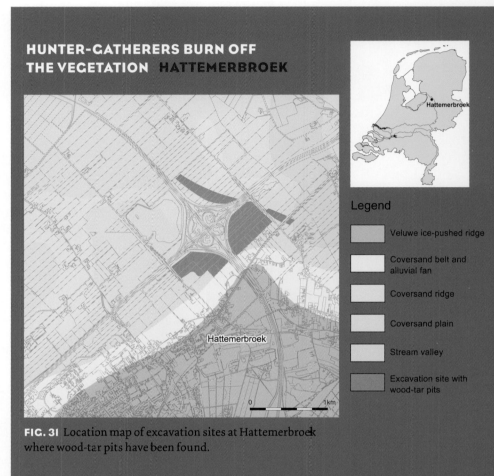

FIG. 31 Location map of excavation sites at Hattemerbroek where wood-tar pits have been found.

During the Mesolithic, generations of hunter-gatherers would go to a wooded sandy ridge along a distributary of the then IJssel river. They went there to make wood tar, a sticky substance that they used as an adhesive, for instance to attach flint arrowheads to wooden shafts. Wood tar was extracted in a process involving the controlled smouldering of wood in pits. Many hundreds of these pits were dug in this area over the centuries.

Thus although the hunter-gatherers were closely connected with their natural environment, they did leave substantial traces behind in the landscape. They dug pits, felled trees and burnt off the vegetation. In this way they created in places a more open vegetation, with a wide diversity of animals. In wetland areas the burning-off of trees led to the growth of young shoots and stalks of soft wood. These were attractive to beavers who helped to create a wetter environment by building dams. And beavers, in turn, were valued by hunter-gatherers for their meat and fur. Thus the presence of hunter-gatherers in this seemingly pristine landscape could be seen – and smelt – in the vegetation.

3850 BCE
EXPANDING PEAT

FIG. 32 A large peat area in Patvin-suo National Park, Finland. Peat areas like this were common in the Netherlands at this time. If trees did grow, they would have been deciduous trees, not the conifers shown in this photo.

By about 3850 BCE the Netherlands had stopped declining in size. Although the sea level continued to rise, the deposition of sand and clay was lifting the mainland at about the same pace. Vast peatlands appeared in both the low and high Netherlands. The people in this area continued to hunt, fish and forage for food in time-honoured fashion. But from 5300 BCE they also became acquainted with agriculture and livestock farming, starting in southern Limburg.

In about 3850 BCE the coastline of the western Netherlands had reached its eastern-most point and flooding from the sea came to an end. The beach barriers and small coastal dunes that formed during this period were no longer swept away by coastal erosion but were preserved. Examples are the small dunes at Ypenburg and Schipluiden at Delft, the oldest surviving coastal dunes from the Holocene.

The mean sea level in this period was 4 to 5 m below NAP, with the rate of sea-level rise dropping to about 30 to 40 cm per century. As a result of the ongoing sea-level rise, the tidal basins and coastal peat zones in Zeeland and the IJsselmeer area had expanded landward. We also see that the tidal basins in the northern Netherlands had grown in size. East of the South Holland tidal basin, peat growth was already intensifying. Because the tidal range along the barrier coast was expanding (from less than 1 m to about 1.8 m) and the tidal basins themselves had grown, this boosted the flow rate in the tidal basins. With much greater volumes of water flowing in and out at each tide, the sea deposited not only clay, but also sand.

In this period the northern Dutch coastal area wasn't yet

3850 BCE

0   50km

Holocene landscape

	Beach barriers and low dunes
	High dunes
	Beach plains and dune valleys

	Tidal flats
	Salt marshes and floodplains
	Salt-marsh ridges and tidal levees
	Peat areas

	Embanked salt marshes and floodplains
	Reclaimed lake
•	Towns and cities
	Urban area

Pleistocene landscape

	Outer water and inner water
	Pleistocene sand areas, below 16 m. -NAP
	Pleistocene sand areas, 16 - 0 m. -NAP
	Pleistocene sand areas, above 0 m NAP

	Floodplains and stream valleys
	River dunes
	Loess area
	Drift-sand areas

	Ice-pushed ridges, ice-pushed till and ridges and valleys shaped by flowing land ice
	Areas with Tertiary and older deposits
NAP	Amsterdam Ordnance Datum

FIG. 33 The earliest
'industry' in the
Netherlands was
the mining and
working of
high-quality flint
that was found
underground in
southern Limburg.
Between 4000 and
3000 BCE countless
tonnes of this
material were
mined. Some of it
was crafted into
axes, which local
farmers sharpened
on this large
whetstone. It can
still be found in the
woods at Slenaken.

connected to the large tidal basin in the former valley of the Oer-Vecht in North Holland and Flevoland. These areas were separated by a high Pleistocene ridge that ran from Texel and Wieringen (North Holland) to Gaasterland (southern Friesland).

The map shows that the peatlands between the northern Netherlands and North Holland-Flevoland had expanded significantly between 5500 and 3850 BCE. This was due to a combination of the rising groundwater table (which was related to sea-level rise) and the deterioration of local drainage systems.

Also in other parts of the high Netherlands (the Peel, parts of Overijssel, and southeast Drenthe and Groningen) the peat gradually expanded as a result of deteriorating drainage. This was a self-reinforcing process: the more peat was formed, the poorer the drainage became. The peat acted like a sponge that absorbed the rainwater. Since the central high parts of the peatland were fed only by rainwater, these bogs were poor in

FIG. 34 The first farmers from the Linearbandkeramik culture built remarkably large wooden houses with deep-sunk posts. They provided a true home base for each family. This example, measuring 31.5 m in length, was excavated in Stein.

nutrients (oligotrophic). In the lower-lying peat bogs in the coastal zone, which were still regularly flooded by nutrient-rich water, it was predominantly reed-sedge peat that formed.

In the Rhine-Meuse delta the continuing sea-level rise caused the freshwater tidal zone to expand in an upstream direction. By about 3850 BCE an enormous peatland had developed as far as the line linking Utrecht and Den Bosch. Interestingly, the delta also extended in a downstream direction between 5500 and 3850 BCE. This seaward expansion was possible because the Rhine and Meuse had supplied so much sediment to the delta that it offset the effect of the sea-level rise. The delta filled not only with river sand and clay, but also large quantities of often clayey wood peat that had formed in the floodplains. Thus the river delta was for the most part filled with peat, which was itself made up of 60 to 90% water.

In the meantime, the rivers in this area were constantly

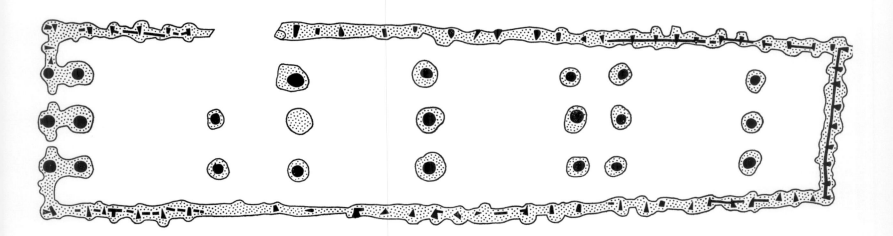

changing course, a process that would continue into the Middle Ages. The shifting of the main Rhine channel had a major impact. Until about 4000 BCE the main drainage channel flowed through the central part of the delta, but for about the next 4000 years it would flow much further to the north. This watercourse is still visible today as the Kromme Rijn, Leidsche Rijn and Oude Rijn.

In about 5300 BCE, when Mesolithic hunter-gatherers were still active in the increasingly water-logged landscape of the coastal region, the first farmers appeared in the southern Limburg loess region. They belonged to the 'Linearbandkeramik' (LBK) culture, named after the bands of incised lines with which they decorated their pottery. This oldest culture from the new Stone Age (Neolithic) had its origins in the Danube region. From there, it spread out across Europe, particularly in areas where the immediate subsurface consisted of fertile loess soils. In the dense forest that covered these soils, these people created clearings for fields and they founded settlements, sometimes palisaded, containing several sturdy houses.

These farmers brought with them food crops and livestock: emmer and einkorn wheat, peas, lentils, linseed, poppy seed, cattle, sheep, goats and pigs. Local deforestation caused erosion on the fringes of the loess plateaus, but only on a fairly limited scale as yet. The presence of these agrarian communities was of short duration (until about 5000 BCE).

Research in Germany and Belgium has shown that violent conflicts occurred.

A tradition of hunting and gathering continued in the low-lying parts of the Netherlands. From 5000 to 3400 BCE, the Rhine and Meuse deltas, the levees of the Rhine, the catchment area of the Oer-Vecht and the predecessors of the Gelderse IJssel, as well as stream valleys, peat bogs and beach zones, were inhabited and exploited by people from the Swifterbant culture (named after the first find site in the Flevopolder). Like their Mesolithic predecessors, these people had hunting, fishing and food-gathering as their main means of subsistence. Of necessity, they were well attuned to the dynamics of the landscape and were able to take full advantage of the presence of red deer, wild boar, beavers and otters, as well as the wide range of birds, fish and edible wild plants. They gradually expanded their options by keeping some livestock: cattle, sheep, goats and pigs. They also engaged in small-scale arable faming (horticulture is a more appropriate term) on the low levees, where they cultivated hulless (or 'naked') barley and emmer wheat. For a long time, however, the direct impact of these agrarian activities on the landscape was limited.

All parts of the landscape were exploited during this period. Flint for toolmaking was collected from glacial till outcrops and beaches. However, the landscape was utilised for more than just food gathering. Boggy areas in particular played a role in rituals, with a range of objects – such as flint tools, antlers and earthenware jars – being deposited in bogs. This custom of deliberately depositing objects in bogs would continue well into the Middle Ages.

So far we know little about habitation in the more elevated parts of the Netherlands at the time of the Swifterbant culture. The most striking phenomenon is the exploitation of flint mines in southern Limburg between 4200 and 3400 BCE.

Flint was mined from shafts and tunnels at Rijckholt, Valkenburg and elsewhere. Mining for flintstone nodules was associated with the specialist production of flint axes and long 'daggers' that were exchanged over long distances – a phenomenon that we also observe elsewhere in Europe during this period.

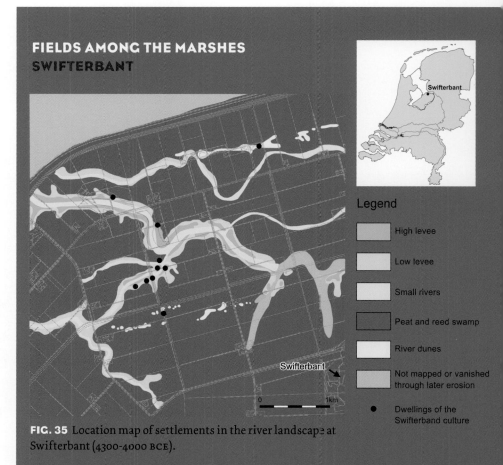

**FIELDS AMONG THE MARSHES
SWIFTERBANT**

Legend

High levee

Low levee

Small rivers

Peat and reed swamp

River dunes

Not mapped or vanished through later erosion

● Dwellings of the Swifterband culture

FIG. 35 Location map of settlements in the river landscape at Swifterbant (4300-4000 BCE).

Between 4300 and 4000 BCE there was a river landscape near Swifterbant, with low, clayey levees, basin clay soils and *donken* (small sand dunes). Marshy conditions prevailed and, since the coast wasn't far away, the groundwater table fluctuated with the tides.

This landscape was inhabited by people from the Swifterbant culture, who lived mainly by hunting, fishing and foraging for food and other natural resources. In that respect they perhaps differed little from their Mesolithic ancestors, just two millennia earlier. Like them, they didn't live in permanent settlements either.

Yet there was a significant difference. They had succeeded in adding to the already-substantial food supply in this marshy environment. On the levees that were scattered across the landscape, they cultivated old varieties of our present-day cereal species: emmer wheat and hulless barley. And they worked their fields with simple hoes.

The fact that they lived in a dynamic landscape was more of an advantage than a disadvantage. At high tide the fields were flooded and fertile clay was deposited. The Swifterbant people had to put up with the fact that flooding wasn't always confined to the fields on the levees. The higher, and therefore drier, *donken* where they lived and buried their dead were also inundated from time to time.

2750 BCE
THE COASTLINE CLOSES

FIG. 36 Pioneer dunes on the island of Schouwen. Here we see the beginnings of dune formation on a wide beach that will continue to widen over time. Although a rare sight along the Dutch coastline today, it was a common phenomenon in about 2750 BCE.

The sea level had risen enormously since the end of the last ice age and for a long time the Dutch coastline looked like an open Wadden landscape. In 3500 BCE this began to change as the coastline closed – at least in the west. In the north it remained open, as it has to this day. Inland, peat growth continued. The farmers who inhabited this varied landscape exploited it more intensively than ever. In the high Netherlands they began burying their dead under distinctive burial mounds.

Between 3500 and 2500 BCE the sea-level rise declined from 30-40 cm to 20-30 cm per century. This was because the ice caps of North America and northern Europe had almost completely melted. The isostatic subsidence of the Netherlands now became the main factor behind the relative rise in sea level, which varied from place to place. In the meantime the sea continued to bring sand and clay to the Dutch coast, in larger quantities than were needed to compensate for the sea-level rise. As a result, the beach barriers began extending seaward and, despite a continuing sea-level rise, accretion and peat growth began in the coastal area behind them. The tidal basins silted up and shrank in size, and the salt marshes expanded. This led to a reduction in tidal volume (basin storage capacity) in the western Netherlands tidal basins, which in turn sharply reduced the tidal channels and inlets. Subsequently, the beach barriers along the coast began to join up more and more, the tidal inlets became even smaller, and less sea water flowed in and out of the tidal basins. Eventually the coast closed almost completely. The closed coast shielded the tidal basins from the sea. As a result, the western Netherlands tidal

48

2750 BCE

Holocene landscape

Beach barriers and low dunes

High dunes

Beach plains and dune valleys

Tidal flats

Salt marshes and floodplains

Salt-marsh ridges and tidal levees

Peat areas

Embanked salt marshes and floodplains

Reclaimed lake

• Towns and cities

Urban area

Pleistocene landscape

Outer water and inner water

Pleistocene sand areas, below 16 m. -NAP

Pleistocene sand areas, 16 - 0 m. -NAP

Pleistocene sand areas, above 0 m NAP

Floodplains and stream valleys

River dunes

Loess area

Drift-sand areas

Ice-pushed ridges, ice-pushed till and ridges and valleys shaped by flowing land ice

Areas with Tertiary and older deposits

NAP Amsterdam Ordnance Datum

0   50km

FIG. 37 Besides arable and livestock farming, hunting and fishing long remained important means of subsistence in the wet western Netherlands. This basket trap, woven from small branches, was found at Vlaardingen and is a product of the culture of the same name. It dates from the period 2900-2600 BCE.

basins changed from saltwater to fresh water and peat began forming on an immense scale.

By about 2750 BCE the coastline of the western Netherlands was virtually closed. Low dunes formed on the beach barriers, interspersed with low-lying beach plains. Only where larger rivers discharged into the sea were there still openings in the coastline: the West Frisian tidal inlet at Bergen, the Oer-IJ at Heemskerk, the Oude Rijn at Katwijk and the Maasmond at Hoek van Holland. The coast of Zeeland remained open.

New peatlands formed behind the closed coastline. Now that the closed tidal basins were deprived of fresh sea water, the water changed into fresh water and the basins evolved into extensive fields of reed. The dead reed sank into the water and turned into reed peat. Over time these connected peatlands expanded inland. By about 2750 BCE the tidal basin in the former valley of the Overijsselse Oer-Vecht in North Holland and Flevoland had largely silted up and become overlain with peat. Only the northwestern part was still a remnant of the tidal area, the West Frisian tidal inlet. There were large lakes in the peat in the present-day IJsselmeer region.

Meanwhile something different was happening in the northern Netherlands: the sea continued to flood the coast,

causing the tidal basins of the Boorne, Hunze and Fivel to even increase in size. This meant that, unlike in the western Netherlands, the coastline of the northern Netherlands stayed open. Throughout the Holocene, this area featured Wadden islands separated by large tidal inlets. There were three factors behind this:
- The subsidence of the northern Netherlands was greater than in the central and southwestern Netherlands, which meant that more sediment was needed to compensate for the sea-level rise.
- Because of the prevailing westerly wind, wave action brought much more sand to the western coastline (perpendicular to the prevailing wind) than to the northern coastline (parallel to the prevailing wind).
- In the western Netherlands this larger volume of transported sand was supplemented still further by sand that was brought to the sea via the Rhine and to a lesser extent the Meuse.

Because the coastline in the northern Netherlands remained open, it has always been subject to strong storm surges and tidal ebb and flow.

Peat continued to form in the high Netherlands. These peat bogs joined up with the coastal peat bogs and by 2750 BCE vast tracts of peatland had evolved between western Brabant and northern Drenthe. The central parts of this peatland featured 'oligotrophic' raised bogs, which grew to several metres above their surroundings and which relied on rain for their water supply. Peat streams took the surplus rainwater from the peat domes to the lower parts of the landscape, where nutrient-rich ('eutrophic') wood and reed peat could develop, or moderately nutrient-rich ('mesotrophic') sedge peat. Where the surface water stagnated, shallow lakes arose, as in Flevoland and the IJsselmeer region.

The changes in the Rhine and Meuse freshwater tidal zone were fairly minor. The rivers changed course occasionally, but the main channels of the Rhine (in the northern part of the delta) and the Meuse (in the south) stayed more or less in place. However, the river system contained fewer branches in 2750 BCE than previously because of the smaller rise in sea level. The northern main course of the Rhine transported so much

FIG. 38 Reconstruction of a farmhouse belonging to the Funnelbeaker culture (Hunebedcentrum, Borger). Megalithic tombs (*hunebedden*) represented the 'home' of the ancestors of this culture. Only a small proportion of the deceased were interred in the *hunebedden*.

sediment that a delta began to form at the river mouth at Katwijk. This phenomenon didn't occur at the Maasmond because the Meuse carried less sediment from the hinterland. There was even a small tidal basin there.

In the eleven centuries from 3850 to 2750 BCE the Netherlands was inhabited by farming communities of the Funnelbeaker, single grave and Bell Beaker cultures. Because their dwellings are barely visible archaeologically, the presence of these cultures can be difficult to establish. Their way of life has largely been reconstructed on the basis of settlement waste and burials, which reveal that they used the landscape in different ways.

The wet parts of the landscape weren't suitable for habitation, agriculture or livestock farming. The increase in peat growth therefore meant a considerable reduction in the habitable land and people were obliged to keep moving towards the high Netherlands. At the same time, however, these wet areas were a rich food source, with birds and mammals aplenty to be hunted and fished. There was also a wide variety of edible foods to be gathered, such as shellfish, hazelnuts and other nuts, which were an important supplement to the products from agriculture and livestock farming. The transitional zone between wet and dry areas was therefore a popular place in which to settle. The more elevated parts of northern North Holland, as well as the freshwater tidal zone and the tidal areas along the coast are all examples of such zones.

Agriculture and livestock farming were practised on the dryer, more elevated parts. In the central Netherlands river landscape, these now included river levees, in addition to the river dunes. The slightly higher abandoned river bed ridges were a peaceful, although changing, environment for a farming existence.

On the sandy soils of the southern and eastern Netherlands the farmers made noticeable interventions in the landscape as they felled parts of the primeval linden and oak forests. They used the timber to build farmhouses, for fuel and as construction timber for roads into the peatland. Various roads made of tree trunks (knuppelpaden) or wooden planks are known from the Drenthe and Groningen part of the Bourtanger Moor. A significant proportion of the oak forest must have been felled and used as road surfaces, leaving the forest much more open.

Fields for the cultivation of hulless barley were created in clearings in the forested sandy landscape. The ard (a simple kind of plough) was introduced and cattle were used as draught animals. On the abandoned fields and other open places where sheep and goats were grazed, the soil became depleted and heathland developed. Farmers grazed their cattle on the pastureland and kept pigs in the oak forests. They also hunted, primarily for red deer and wild boar.

The inhabitants of the high Netherlands buried their dead in forest clearings. They would erect a mound of heather and woodland sods over the grave. The families of the deceased had clearly-defined ideas about the funerary ritual, with only adults being buried beneath the mounds. The dead weren't cremated, but were interred in a curled position, with a standard set of grave goods: a beaker containing a beverage or food and a stone or flint tool. The burial mounds were easily recognisable in the landscape. Later inhabitants sometimes chose to build new burial mounds in the vicinity of the old ones,

giving rise to the first groups of these mounds.

The use of special materials reached a peak during this period as people made increasing use of amber (and very occasionally jet) as a raw material for bead, button and pendant jewellery. Amber occurred naturally in till outcrops and on the northern coast, while jet came from further afield, from the northern coast of France. A new phenomenon was the sparing use of metal (copper and gold), which must have been imported in its raw state or as ingots from different parts of Europe. Copper and gold are easy-to-work metals that could be hammered into the desired shape. Their nearest source regions were western France, northern Spain and the British Isles. The fact that farmers in the Netherlands had access to gold and copper shows that they exchanged goods over long distances.

FIG. 39 Location map of the grave at Sijbekarspel.

She had been placed very carefully in the burial pit at the end of the late Stone Age. Her family had buried her in the traditional way: lying curled up on her left side with her face to the south.

Her skeleton – even her dental plaque – was exceptionally well preserved in the chalky, clayey soil. An analysis of the plaque showed that she had eaten food that had come from a saline or brackish environment, as well as cereals. This is very much in keeping with findings from other studies. She had lived in a vast saltmarsh landscape, intersected by creeks and gullies.

This woman belonged to a late Neolithic farming community that made smart use of the surrounding landscape. They cultivated barley and wheat in elevated areas and regularly helped themselves to mussels from the mussel beds, which were dry at low tide. In the dry season the salt marshes produced grass in abundance, ideal for the grazing of cattle, sheep and goats. Wild duck and pike were also on the menu.

Some time after her burial, the landscape became increasingly wetter. Peat starting to grow and after a few centuries the settlement and its grave disappeared from view – until 1989 that is, when archaeologists began their excavations in Sijbekarspel.

1500 BCE
PEAT COVERS THE LAND

FIG. 40 Shallow lakes in vast reed fields were common in the Rhine-Meuse delta until about this time. Later, most of them became overgrown with peat. This photo was taken in the Danube delta in Romania.

From 3500 BCE onward, rows of beach barriers with low dunes evolved along the entire west coast and the coastline became increasingly closed. Despite the ongoing sea-level rise, the sediment supply was so great that the coastline had even shifted several kilometres seaward. Between the beach barriers and the elevated Pleistocene land, peat formation continued unabated and by about 1500 BCE half of the Netherlands was covered with peat. On the dry Pleistocene soils, humans tightened their grip on the landscape: forests increasingly gave way to fields and pastureland, and for the first time small hamlets made their appearance.

In around 1500 BCE the sea-level rise had declined to about 15-25 cm per century. Along the coasts of Zeeland and Holland the beach barriers and dunes had shifted further seawards and a long barrier coast had evolved. This was due to the large quantities of sand brought to the coast by currents and surf: it meant that the beaches were elevated at a faster pace than the sea level was rising. The only places where the coastline was interrupted were the West Frisian tidal inlet at Bergen, the Oer-IJ at Castricum and the mouths of the Oude Rijn, Meuse and Scheldt. Via these inlets and rivers, the hinterland drained into the North Sea. Marine clay was only deposited in the estuary areas of the tidal inlets and river channels. The sea exerted much less influence than in 2750 BCE, especially in Zeeland and western Friesland.

Behind the beach barriers the entire western Dutch coastal landscape had turned into one vast peat bog. Large lakes had appeared in the IJsselmeer region, which continued to expand over time as wave action caused their shores to crumble away.

The open tidal area in the northern Netherlands – with its islands and tidal inlets off the coast – persisted. This was

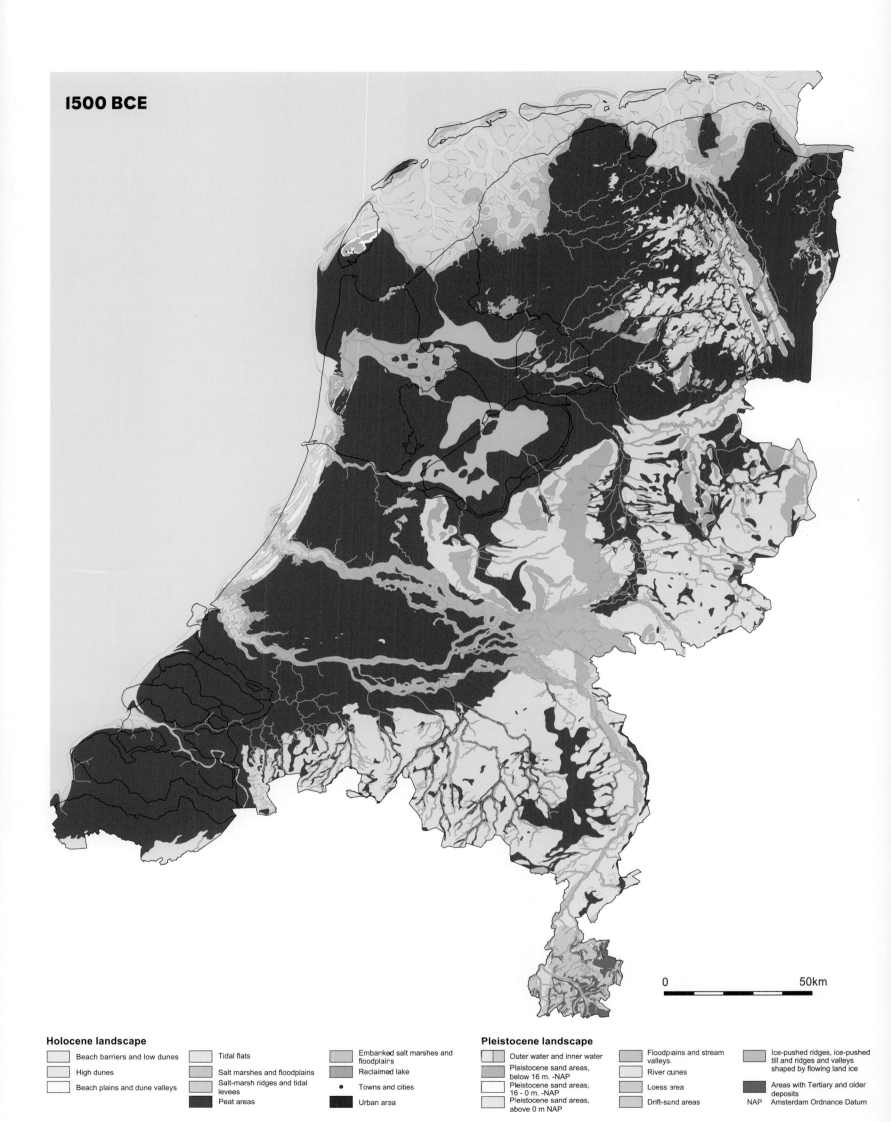

1500 BCE

Holocene landscape

Beach barriers and low dunes

High dunes

Beach plains and dune valleys

Tidal flats

Salt marshes and floodplains

Salt-marsh ridges and tidal levees

Peat areas

Embanked salt marshes and floodplains

Reclaimed lake

Towns and cities

Urban area

Pleistocene landscape

Outer water and inner water

Pleistocene sand areas, below 16 m. -NAP

Pleistocene sand areas, 16 - 0 m. -NAP

Pleistocene sand areas, above 0 m NAP

Floodplains and stream valleys

River dunes

Loess area

Drift-sand areas

Ice-pushed ridges, ice-pushed till and ridges and valleys shaped by flowing land ice

Areas with Tertiary and older deposits

NAP   Amsterdam Ordnance Datum

0   50km

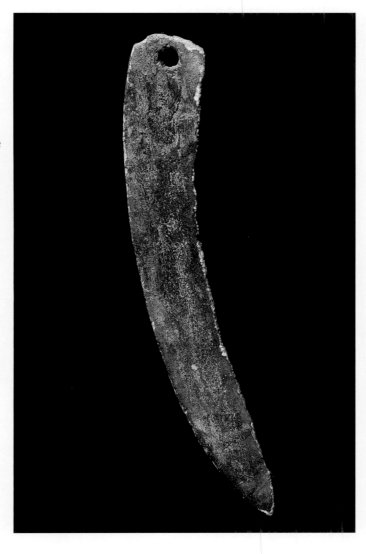

edges of the northern Netherlands tidal basins (the Boorne, Hunze and Fivel) did accretion occur. There, the peat expanded over the tidal deposits in a seaward direction. However, there was no large-scale accretion and overlying of peat – as happened in the western Netherlands. The location of the Wadden islands at this time is uncertain. We do know, however, that they were situated slightly further to the north than the present-day Wadden islands.

In the river area, several branches of the Rhine had since disappeared and become overgrown with peat. New branches had emerged which discharged to the mouth of the Meuse (Maasmond). River clay was deposited along the active rivers, while peat formed in the lower-lying areas further away from the rivers. The fact that most of the water and sediment from the Rhine was discharged through the Oude Rijn is evident in the coastline at Rijnmond. A small delta, made up of sand and clay supplied by the Rhine, had developed where the river flowed into the North Sea. Because the Meuse supplied much less river sand, the coast couldn't extend in a seaward direction and the mouth of the Meuse was subject to greater influence from the sea. Compared with the Rhine branches and the Meuse, the Scheldt was only a small river that supplied very little sediment and was completely imbedded in the peatland.

In the high Netherlands the peat continued to expand. In the northern Netherlands in particular, the stream valleys were increasingly filled with peat, causing further deterioration in the drainage of the high Netherlands. The drainage was also worsening elsewhere in the high Netherlands, causing the peatlands to expand still further.

largely due to the prevailing westerly direction of the sea current and the wind, which meant that the sediment supplied by the sea could be washed away again (unlike in the western Netherlands, where the prevailing wind came at right angles to the coast). In addition, there were no major rivers to supply sand. All in all, this meant that there wasn't enough sedimentation to compensate for the rising sea level. Only along the

Although these circumstances – a growing expanse of marshy peat – don't appear conducive to habitation, the people who occupied what is today the Netherlands managed to survive remarkably well. Wherever it was possible to live, they did so, and the population density increased in those habitable areas. The west Frisian salt-marsh landscape and the central Dutch river area were densely populated, whereas the peatlands in

the north and behind the southwestern beach barriers appear to have been scarcely inhabited at all.

Initially the famers lived in scattered farmsteads, a tradition continued for a long time by farmers on the sandy soils of the southern and eastern Netherlands. But elsewhere farms moved closer together and it gradually became more common for people to live in small hamlets. This occurred, for example, in the river area and on the west Frisian creek ridges. Around these hamlets the landscape was partitioned by ditches and wicker fences. The farmsteads were laid out with a variety of outbuildings, such as granaries in which the harvest was stored. People and livestock sought shelter in 'byre houses', where they lived under one roof – a tradition that lasted well into the 20th century CE.

The inhabitants of these farming settlements made ever-greater interventions in the landscape. Especially in regions of intensive habitation, such as the higher sandy soils and the river area, the landscape acquired a more open appearance because of large-scale tree-felling. This was probably how beech trees came to be established in our part of the world, in around 2000 BCE. Arable and livestock farming developed into a single integrated (or mixed) form of farming, whereby farmers used animal manure to fertilise their fields. They may also have alternated the cultivation of different crops so that the soil wouldn't become too depleted (crop rotation). Innovations of this kind made a still higher population density possible.

Livestock farming (especially cattle, and to a lesser extent pigs, sheep and goats) clearly ensured a stable supply of animals for food, since these farmers scarcely engaged in hunting any longer. Wild animals (elk, red deer, roe deer, wild boar and fur animals) constituted only a minor supplement to their diet. We see a similar phenomenon in the cultivation of crops: wild plants were now a negligible part of the diet, whereas barley and wheat had become important food sources.

People also started to dress differently in this period. The presence of loom weights indicates that sheep were also kept for their wool and that clothing was no longer made solely from animal skins. Also new was the use of horses as beasts of burden and mounts. We know that flax was cultivated in this period, which means that these people were perhaps already making linen clothes. We can't be entirely sure about this, however; flax seeds may have been grown for their oil.

The major changes in the lifestyle of these farmers went hand in hand with innovations in rituals surrounding death. Burial mounds continued to be erected, but no longer for a single deceased individual. New graves were made in the flanks of existing burial monuments, whereby burial mounds became places of burial for several individuals. Cremation also became popular during this period.

One of the most conspicuous changes (for archaeologists) at this time was the gradual disappearance of flint as the raw material used for cutting tools such as knives and axes. Increasingly, people turned to metal – initially copper, and later bronze (an alloy of copper and tin). These raw materials were not available in the Netherlands. Instead, Dutch farmers acquired their bronze items through vast exchange networks with all corners of Europe. These networks had been around for quite some time but the advent of bronze may have changed

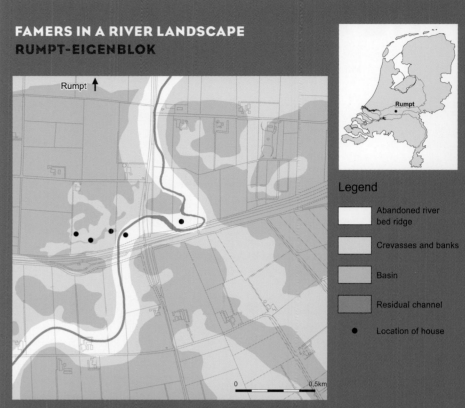

FAMERS IN A RIVER LANDSCAPE RUMPT-EIGENBLOK

Rumpt

Legend

Abandoned river bed ridge

Crevasses and banks

Basin

Residual channel

● Location of house

FIG. 43 Location map of Bronze Age farmsteads at Rumpt.

During the Bronze Age a small settlement was located in the river area near present-day Rumpt. It consisted of a few long 'byre houses', in which people and livestock lived under one roof. Fences separated the fields, pastures and the inhabited parts of the settlement, which featured various sheds and storerooms. The area was used intensively: the farmhouses were only 100 to 300 metres apart.

The reason why these farmers lived so closely together relates to the landscape. Fields that are too boggy can't be used and there were no dykes at that time. Farmers therefore had to rely on natural elevations in the landscape, such as former river banks and 'crevasse deposits',

places where the river had breached a levee, leaving behind large quantities of sand. Somewhat higher than the surrounding landscape, fertile and well-draining, these fickle and at times narrow elevations were good places to live and to grow crops.

The Bronze Age settlement at Rumpt was situated on crevasse deposits of this kind. On the periphery were wet grasslands where livestock grazed, as well as wet woodland, reed beds and small ponds. Farmers could therefore till their fields, graze livestock and catch fish within easy reach of home. They may have used canoes to maintain contacts with people in other areas.

both their focus (on source regions for raw materials) and significance. Controlling the flow of bronze and organising forms of recompense could suggest the presence of regional 'leaders'. Some of the bronze items that came in via these exchange networks were transformed into regional bronze products. Others – including personal possessions such as swords, spearheads, jewellery and axes – would be ritually deposited in wet locations, such as streams, pools and bogs. These rituals may have marked the transition to a new stage of life for the former owner.

500 BCE
THE BUILDING OF TERPS

FIG. 44 A flooded salt marsh on the Dutch island of Schiermonnikoog. Salt marshes are located above the mean high water level and are only inundated during spring tides and storm surges. These days they are rare, but in about 500 BCE large parts of Groningen and Friesland looked like this.

In 500 BCE the Netherlands looked very much the same as it had 1000 years earlier. The West Frisian tidal inlet had silted up, however, and peat growth was further advanced. In the eastern Netherlands river area the peat bogs were overlain with river clay, the result of human intervention. Massive deforestation in Germany meant that the Rhine was supplying increasing quantities of clay. Other changes in the Dutch landscape were also the result of human intervention: the last substantial vestiges of primeval forest were felled and for the first time people began building dwelling mounds (terps) in the northern Wadden region.

The coast of the western Netherlands had hardly changed at all between 1500 and 500 BCE. The sea-level rise had declined markedly and currents and wave action had supplied so much sand that the coastline had shifted westward. The West Frisian tidal inlet, one of the last such inlets, was now completely sanded up and cut off from the sea by beach barriers. As a result, by about 1000 BCE the area could no longer drain effectively via natural channels. Like other areas behind the coastline, this former tidal zone became increasingly wet and was eventually completely covered with peat.

Several openings remained in the western Netherlands coastal barrier. One of them was the Oer-IJ mouth at Castricum, where the waters of the lake region at Almere discharged into the sea. Some of this water came from the Rhine, since Rhine water also flowed via the Utrechtse Vecht into the Flevomeer, the region's southern lake. This greater discharge may have widened the Oer-IJ, thereby increasing the tidal ebb and flow in that area. It could also explain why new creek systems developed in the peat area around the Oer-IJ.

In the northern Netherlands the tidal basins of the Boorne,

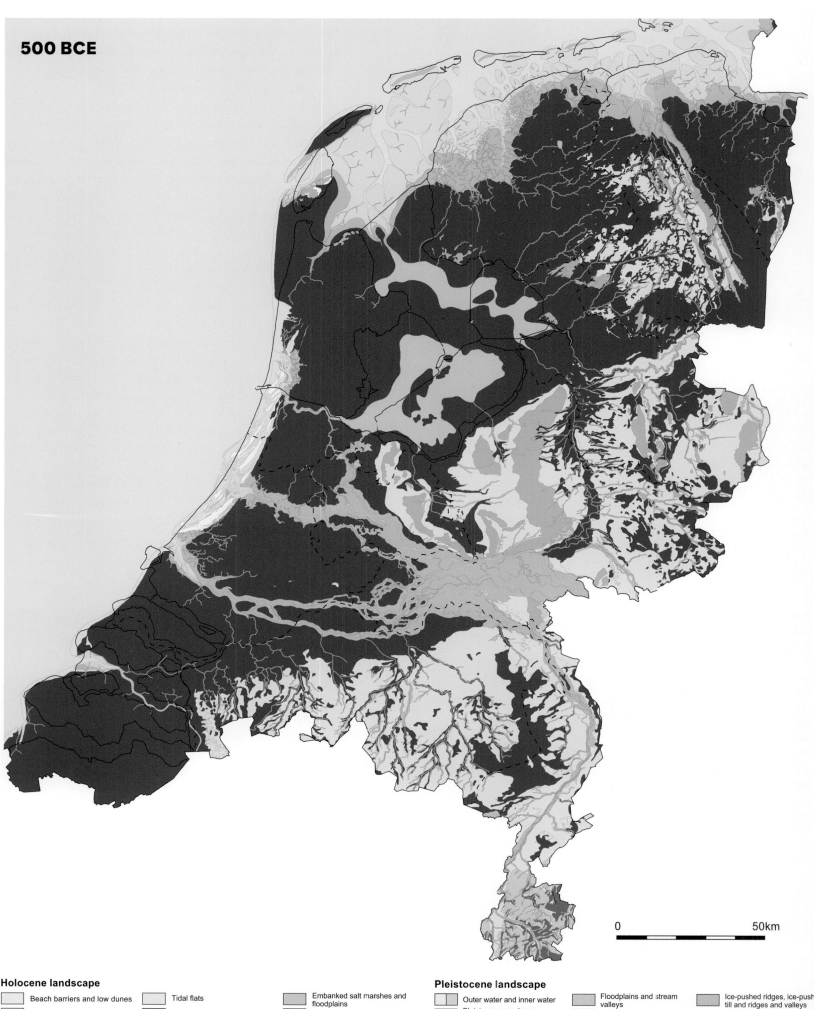

500 BCE

Holocene landscape

	Beach barriers and low dunes
	High dunes
	Beach plains and dune valleys
	Tidal flats
	Salt marshes and floodplains
	Salt-marsh ridges and tidal levees
	Peat areas
	Embanked salt marshes and floodplains
	Reclaimed lake
•	Towns and cities
	Urban area

Pleistocene landscape

	Outer water and inner water
	Pleistocene sand areas, below 16 m. -NAP
	Pleistocene sand areas, 16 - 0 m. -NAP
	Pleistocene sand areas, above 0 m NAP
	Floodplains and stream valleys
	River dunes
	Loess area
	Drift-sand areas
	Ice-pushed ridges, ice-pushed till and ridges and valleys shaped by flowing land ice
	Areas with Tertiary and older deposits
NAP	Amsterdam Ordnance Datum

0   50km

FIG. 45 In the Middle Bronze Age (500-250 BCE), in a large peat area where modern-day Spijkenisse is located, someone stuck this ash-wood spade upright in the ground and left it there. This agricultural implement was found intact by archaeologists.

drowned earlier (1000-750 BCE) than those of eastern Oostergo (500-400 BCE).

The map shows the expansion of the peat lakes of the IJsselmeer region. The lake boundaries are not certain, however. This reconstruction assumes that the northern lake, into which the Overijsselse Vecht flowed, discharged into the Wadden Sea. That too is uncertain. The large southern Flevo lake was connected to the sea via the Oer-IJ.

The peat areas of the coastal marshes and the high Netherlands had grown in relation to 1500 BCE, creating an enormous peatland that covered over half of the Dutch territory. Raised domes of peat moss had formed in both the coastal zone and the high Netherlands. These raised bogs kept growing, sometimes to heights of several metres above the adjacent streams and rivers. It was primarily reed-sedge peat that formed along the fringes of the tidal basins and inlets, and mainly wood peat in the freshwater tidal zone of the Rhine and Meuse. The peat areas in Overijssel, the Peel and western Brabant also continued to expand.

The most obvious change in the river area is the significant increase in clay deposits in the east. This clay layer was formed as a result of a significant increase in sediment supplied by the Rhine. This in turn was due to the intensive deforestation of the German hinterland, where people had transformed forested areas into fields. Heavy rainfall in autumn and winter could now easily wash away the soil, which would find its way via streams and tributaries to the Rhine. In times of flood, the Rhine then deposited this eroded soil as clay in the flood basins alongside the rivers.

The rivers continued to change course during this period. Although a growing number of Rhine branches flowed into the North Sea via the Maasmond, the Oude Rijn remained the main branch of the Rhine. The supply of sediment via that river course was such that the river delta continued to extend into the sea at Katwijk.

Hunze and Fivel still existed but peat expansion came to an end in the northern-most peatlands of Oostergo (Friesland) and the Wold region (Groningen). Between 1000 and 400 BCE these areas were inundated once again and covered with a layer of clay. This flooding didn't occur everywhere at the same time. The peatlands of western Oostergo, for example,

FIG. 46 The salt marshes of Groningen and Friesland became inhabited during the Early Iron Age. The remains of the first timber farmhouses were later covered over when terps were raised, ensuring their preservation, as was the case here in Ezinge. This terp was excavated in the 1930s.

In the Maasmond area, the main channel migrated in a south-westerly direction towards Voorne, causing the tidal creeks northeast of the channel (the Oer-Gaag system at Delfland and Maasland) to silt up. The resulting drop in natural drainage enabled the peat to expand across older salt-marsh deposits.

The Scheldt was still a fairly small river in a large peat area.

In the northern Netherlands, the habitable area had again declined significantly over a period of a thousand years as peat continued to overlay the Drenthe Plateau. The neighbouring tidal zone was not yet inhabited during the Bronze Age. As the sea level rose less rapidly, the salt-marsh area had expanded considerably in the Bronze Age. It wasn't until the early Iron Age that people settled once and for all in the salt marshes, especially on the highly silted-up salt-marsh ridges and levees, which were not inundated for long periods of time. In places where the salt marsh flooded during storms, people built dwelling mounds (*terpen* or *wierden*) by stacking salt-marsh sods on top of one another. This created an extensive terp landscape that was mainly used as pasture for livestock.

In western Friesland the landscape grew wetter as the West Frisian tidal inlet silted up. In the late Bronze Age, the steadily expanding peatland forced the inhabitants to seek refuge elsewhere. Only a few locations that could drain to the Flevo lakes via watercourses in the peat remained habitable in the eastern part of West Friesland. Where this wasn't possible, people moved towards the coast – in this case to the west, where they settled on the beach barriers. These were high enough that there was no need to build terps.

Although the landscape south of the Rhine mouth (Rijnmond) wasn't too wet for habitation, the ground was too waterlogged to grow crops. The inhabitants therefore concentrated on livestock farming. They grazed their animals on the natural grasslands, but also exploited the landscape in another way: in summer, they would let seawater evaporate and would collect the salt that was left behind. Salt was a sought-after commodity that was traded far into the hinterland.

More and more land was put into cultivation on the sandy soils of the southern and eastern Netherlands. The last vestiges of primeval forest disappeared and the heath advanced still further. The farmers would build and rebuild their byre houses anew each time at some distance from the previous one, causing farmsteads to 'wander' across the landscape. The land under cultivation consisted of small square or rectangular plots enclosed by low embankments. Alongside the traditional barley and wheat, farmers also grew new crops in these 'Celtic fields', such as spelt, camelina, rapeseed and broad beans. Unlike the inhabitants of wetter areas, farmers on the sandy soils engaged in mixed farming, in which livestock were mainly kept for manure production and as draught animals. By-products such as bovine milk and sheep's wool became increasingly important during this period. Apart from hazelnuts, blackberries and acorns, farmers no longer gathered wild fruit (acorns were eaten once the poisonous tannin had been extracted). Hunting was also virtually a thing of the past.

There were many changes in the burial ritual. Burial mounds were opened up to a growing number of deceased individuals, including children. The burned remains of the dead were buried beneath smaller mounds that were erected

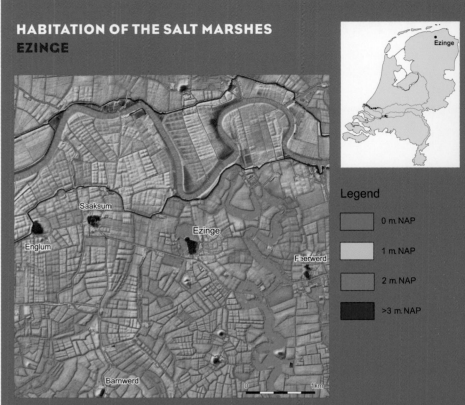

HABITATION OF THE SALT MARSHES EZINGE

Legend

	0 m. NAP
	1 m. NAP
	2 m. NAP
	>3 m. NAP

FIG. 47 The Middag terp region on a LIDAR elevation map.

The Hunze stream valley, set deep in the substratum, played a prominent role in the evolution of the Middag area of Groningen. In about 1500 BCE, the first salt marshes developed in the V-shaped mouth of this river and gradually expanded in a northerly direction over the next one thousand years. By about 500 BCE they had become so silted up and extensive that human habitation was now possible. One of the places where people began to live was Ezinge. We know a good deal about the origins and development of the terp at Ezinge thanks to major excavations by Albert Egges van Giffen in the 1920s and 30s (fig. 46). Habitation started there during the 5th century BCE on what at the time was still a middle salt-marsh zone; in other words, it still subject to regular flooding. The oldest documented features are an enclosed yard with a farmhouse and a large platform supported by rows of posts, which may have been used to protect goods and harvest produce from flooding. During the investigation, more than eighty additional farmhouses and sheds from later periods were excavated. They can be attributed to twenty-nine successive periods of building activity between the Middle Iron Age and the 4th century CE. The terp at Ezinge eventually reached a height of 5.5 m. and a diameter of 450 m.

close together. Many of these large cemeteries, also known as urnfields, have been found on the sandy soils. A rich burial would sometimes mark the beginning of such an urnfield; a sword, axe, knife, horse gear, wagon parts and bronze dishes would be placed in the grave of the 'chief' in question. This is the first time that the burial ritual clearly demonstrates the presence of a warrior elite: men who held higher positions within these small-scale agrarian societies.

Iron was used for the first time in this period, although this didn't mean the end of the local bronze industry. Unlike copper and tin ore, iron ore was available locally in the form of bog iron ore. It was found in swamps, stream valleys and perhaps on the coastal plains. Even so, local production was limited in scale. Most iron objects came to the Low Countries through exchange with regions elsewhere.

250 BCE
CHANGING TIDAL SYSTEMS

FIG. 48 Before dykes were built, the Dutch coast was a succession of large tidal systems. As changes around 250 BCE show, each system evolved differently. To this day, the Wadden Sea is a vast tidal zone, whereas the Slufter area on the island of Texel is of very modest proportions.

Between 3850 and 500 BCE the Dutch coast had become closed to the sea. This process came to a halt after 500 BCE, when the sea was once again able to make inroads into the coastlines of Zeeland and South Holland. Between North Holland and Friesland, a connection opened up to the large lake region of Flevoland (the present-day IJsselmeer). A new tidal system developed in the peat region at Paesens in northeast Friesland. While new tidal inlets appeared, other parts of the Dutch coast accreted. The Oer-IJ estuary was closed off from the sea by a beach barrier and in the northern Netherlands the tidal system of the Boorne accreted and the shallow inlets of the Hunze and Fivel gradually shrank as a result of silting.

In the period around 250 BCE the coastline of Zeeland was gradually broken open. The sand to which the Zeeland coast owed its origins had come from a large Pleistocene sand bank off the Belgian coast, from where it had been shifted by currents and wave action. But this supply of sand ran out around 250 BCE, as a result of which almost no more sand could settle along the Zeeland beaches. In the meantime, the coastal currents continued to be active, carrying sand from the Zeeland beaches to the central part of the Holland coast. Whereas the beach barriers and dunes in the central part of the coastal bight were still expanding, along the Zeeland coast they were washed away and the coastline was eroding. In this period too, the Scheldt was still just a fairly small river in a large peat region.

In the Maasmond area, the sea's influence increased considerably. The low-lying peat lakes near Delft were submerged, enabling the tidal system of the Gantel to expand there. The rise in extremely high water levels in about 250 BCE were caused by a greater discharge of river water from the Rhine towards the Maasmond. When the increased flow from the

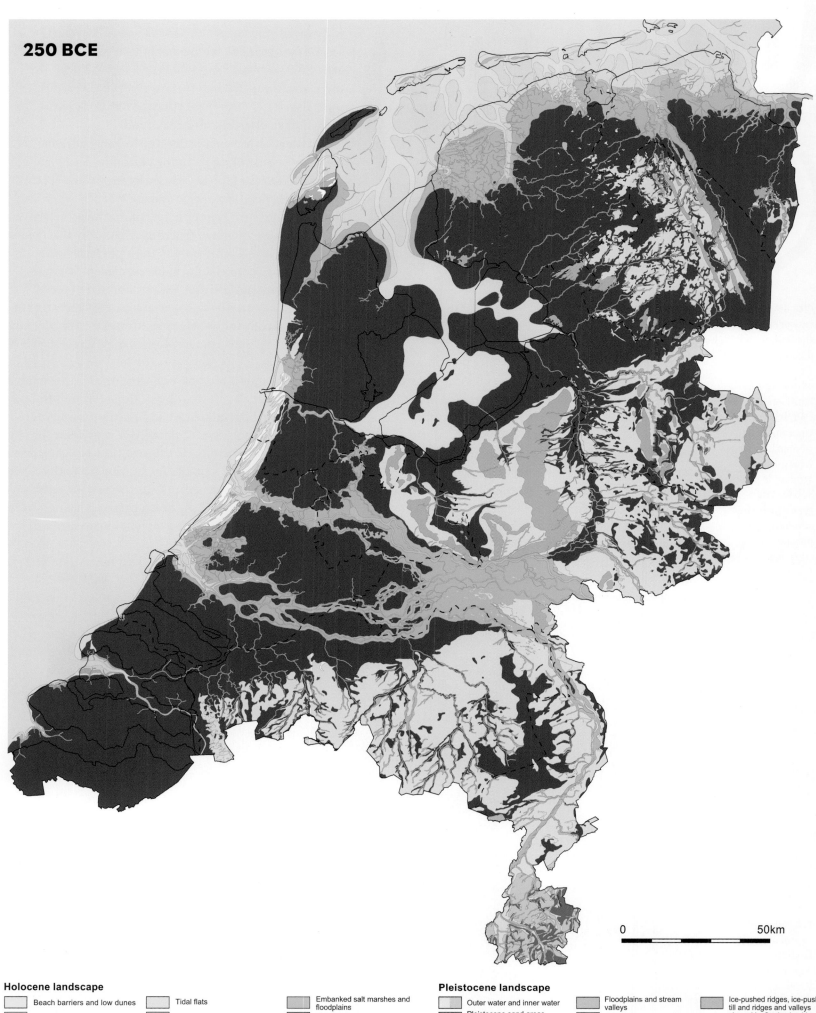

250 BCE

Rhine and the Meuse coincided with higher sea levels, the river water could no longer discharge properly to the sea and the high water levels in the estuary rose enormously.

In Maasland the steady, periodic increases in water level meant that the peatland was no longer flooded but started to float. With extreme flooding, the oxygenated peat soil tore loose from the subsoil. Large islands of floating peat, including the remains of prehistoric settlements that they contained, floated on the water. A clay known as intrusion clay (*klapklei* or *oplichtingsklei*) was deposited beneath the floating peat. Once the water level had fallen again, the floating islands came to rest on this newly-deposited layer of intrusion clay. This alternating floating and sinking of the peat at fluctuating high water levels continued until the peat became so saturated that it could no longer float. After that time (about 200

BCE), the peat was flooded and a layer of clay was deposited on top of the peat islands.

Surprisingly, people had begun to settle in this dynamic landscape around the Maasmond in about 600 BCE. An estimated 25 to 35 people per km² lived there during the third and second centuries BCE, a remarkably high population density. When the peaty soils began to tear loose and float, parts of their farmhouses fell into the peat cracks and became part of the *klapklei* layer. This meant that the remains were well preserved, albeit often in scattered fashion.

The situation around the Oer-IJ also changed dramatically. The river began silting up in the years after about 400 BCE. This accretion was probably caused by an opening appearing between the Flevo lake region and the Wadden Sea, whereby the water from Flevo Lake no longer drained away via the Oer-IJ and the estuary gradually silted up. This meant that the area was inundated less and less often. The higher salt marshes and peat margins became suitable for permanent habitation in the middle and late Iron Age. Excavations in the Broekpolder, on the high salt-marsh ridge west of the Oer-IJ, show that this location was fully inhabited and cultivated from 400 BCE onwards. Farmers grew emmer wheat, barley, flax and gold-of-pleasure.

In about 250 BCE a beach barrier had formed in the mouth of the Oer-IJ, which cut off the estuary almost entirely from the sea. As a result, tidal influence had virtually ceased. After 200 BCE the estuary was cut off from the sea altogether. The sand flats between Castricum and Uitgeest then became permanently dry and thus suitable for human habitation. Only during extremely severe storms was there still a perceptible marine influence behind the beach barrier. The high storm water would flood the lowest parts of the beach barrier, depositing shell-rich sediments there.

Because the coast between the Flevo lake region and the

Waddenzee was broken open, a new tidal zone evolved between Texel and Friesland. The lake region also changed dramatically in response to ongoing peat erosion and increasing tidal influence in the Almere, the predecessor of the Zuiderzee.

In Westergo in Friesland, the former tidal system of the Boorne silted over and became a large salt marsh. The tidal zones of the Hunze and Fivel also shrank as a result of silting. On the other hand, the Middelzee, the tidal system separating Westergo and Oostergo, gradually expanded and the sea forced its way into the peat region at Paesens, in the eastern part of Oostergo. The reason for the Middelzee's expansion was that it took over the drainage function of the Boorne hinterland. This was at the expense of drainage via the course of the former Boorne in Westergo. The sea was able to break through at Paesens into the peat hinterland of northeast Friesland because the protecting beach barrier had disappeared through erosion. The vulnerable peat region now fell under marine influence: tidal channels appeared and a layer of clay was deposited on the peat.

Around 250 BCE, the sea-level rise was 10 cm per hundred years at most, a slight increase that had only a minor impact on the formation and expansion of tidal systems. The driving mechanisms were regional factors, such as the shifting of waterways in the hinterland and the erosion of vulnerable peat areas. Farming communities were able to adapt to these changing circumstances. During times of flooding, the farmhouse floors were the first to be raised, and then later the farmyards themselves. Dams and boardwalks were built to provide access to parts of the landscape that were difficult to reach. Sometimes people withdrew from the peatlands and moved to the dunes. In and between the layers of dune sand, archaeologists have found many traces of fields in the form of plough marks. This suggests that a form of extensive arable farming was practised, involving a series of temporary fields at a different location each time.

In the high Netherlands, little changed. The peat continued to expand until eventually only the highest Pleistocene elements of the landscape, such as the ice-pushed ridges, still remained at ground level. In most cases, the inhabitants of the eastern and southern Netherlands lived in isolated farmsteads within field complexes surrounded by low embankments. Increasingly, they built their farmhouses at the same location. Houses were occasionally clustered together in a small group and some settlements were ditched. Storage areas – for example, at Rhee, Vries and Zeijen – took on the appearance of fortifications. There, on the northern periphery of the Drenthe plateau, an area was established where farmers from the sandy soils could exchange goods such as grain and livestock with farmers from the clay soils.

Salt assumed a growing role in exchanges between regions. Salt production on the coast, for example at Monster, and from mines at Hallstatt in Austria, had to satisfy the growing demand. Salt was used not only as a flavouring agent but also to preserve meat and dairy products. Large quantities of salt were also needed to process animal skins.

Glass was highly sought after too. Brightly coloured glass arm bands found their way into the Netherlands in the late Iron Age. In the first century BCE, they were produced in the Oss region and the eastern river region. These items of jewel-

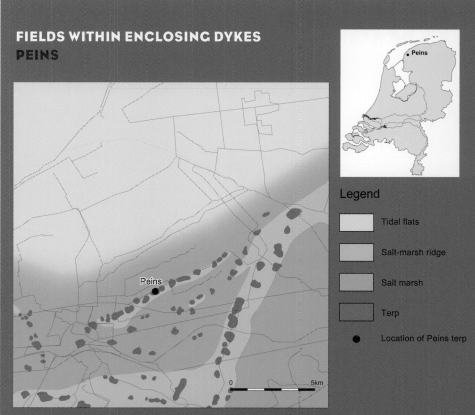

FIELDS WITHIN ENCLOSING DYKES
PEINS

Legend

Tidal flats

Salt-marsh ridge

Salt marsh

Terp

● Location of Peins terp

FIG. 51 Location map of terps in the salt-marsh landscape around Peins at about the beginning of the first century.

Immediately north of the built-up area in the Frisian village of Peins is a terp dating back to the beginning of the Roman period. It is thanks to terps like this one that people were able to inhabit the salt marshes, despite regular flooding by seawater.

But people had been active in this area even before terps were built. In the late Iron Age the inhabitants of terps situated further to the south were making innovative use of the natural accretion of new salt marshes. Initially, they used them as rich grazing ground for their livestock; later, they used the highest ground for cultivation. These fields weren't located in the open countryside but were en-

closed by a low ring dyke.

The dyke at Peins consisted of long salt-marsh sods, which had been dug from the immediate vicinity. Originally the dyke was no more than 1.2 metres high and 3 metres wide. Later, it was extended, mainly in width. The dyke protected the crops from high tides and storm surges in spring and summer. Floods weren't a problem in autumn or winter; people may even have welcomed them as a source of fertile silt.

It wasn't until Roman times that the people exploiting this land decided to live there too. They built small house terps for this purpose, behind or on top of the dyke.

lery were part of the regional attire and probably also a form of currency. They are regarded as the forerunners of the gold and silver coins minted by Celtic tribes.

Little is known about the burial ritual other than that the dead were cremated and their graves were not marked by earth mounds. Sometimes the dead were buried close to or inside a large, square ditched structure, known as a cult place. These were open-air sanctuaries where food offerings were made and ceremonies performed to make contact with the supernatural world. Coins, pottery vessels or other objects deliberately left behind or buried are also viewed as sacrifices. Perhaps the Drenthe bog bodies can also be interpreted in this light.

100 CE

HUMAN IMPACT

FIG. 52 Tidal channel in the 'Drowned Land of Saeftinghe' at Emmadorp, Zeeland. In the first and second centuries CE, the coastal landscape behind the dunes and along the river mouths had subsided as a result of reclamation. This enabled the sea to penetrate the land once again through tidal channels.

The salt marshes and neighbouring peatlands of the Dutch coastal area became inhabited during the Iron Age, and this continued in Roman times. People inhabited and cultivated all the tidal systems from Zeeland to Groningen. Human intervention began to have an impact on the formation of the landscape. Natural drainage was improved by the building of ditches and canals but it also gave the sea easier access to the hinterland and it caused the peat soil to compact and subside. The digging of peat for fuel and for salt extraction also led to subsidence of the peat surface. In the 1st century BCE, when the Oude Rijn marked the border of the Roman empire, the landscape was changed as a result of Roman hydraulic works.

In Zeeland large tracts of the extensive peat region were inhabited and peat was extracted in vast quantities. The large-scale, rectangular parcelling pattern from that time can still be seen on modern elevation maps. In the early Roman period, this exploitation of the peatland had only a limited impact in Zeeland. The upper part of the peat was still relatively high and the sea was not yet able to penetrate deep into the peatland. By about 100 CE the openings in the coast had already grown much bigger. On the island of Walcheren, a salt marsh had developed behind one such opening in the beach barrier and people had taken up residence there. After 270 CE, the land in Zeeland was subject to extensive drowning, which we can attribute to the exploitation of the peatland. Since before the first century CE, people had dug drainage ditches and channels in the peat in order to make it habitable. These waterways drained water from the peatland through the natural openings in the beach barriers. These openings and ditches allowed the sea to penetrate into the heart of the peatland. In addition, the surface of the peat subsided as drainage caused the soil to compact. People also began digging up the peat.

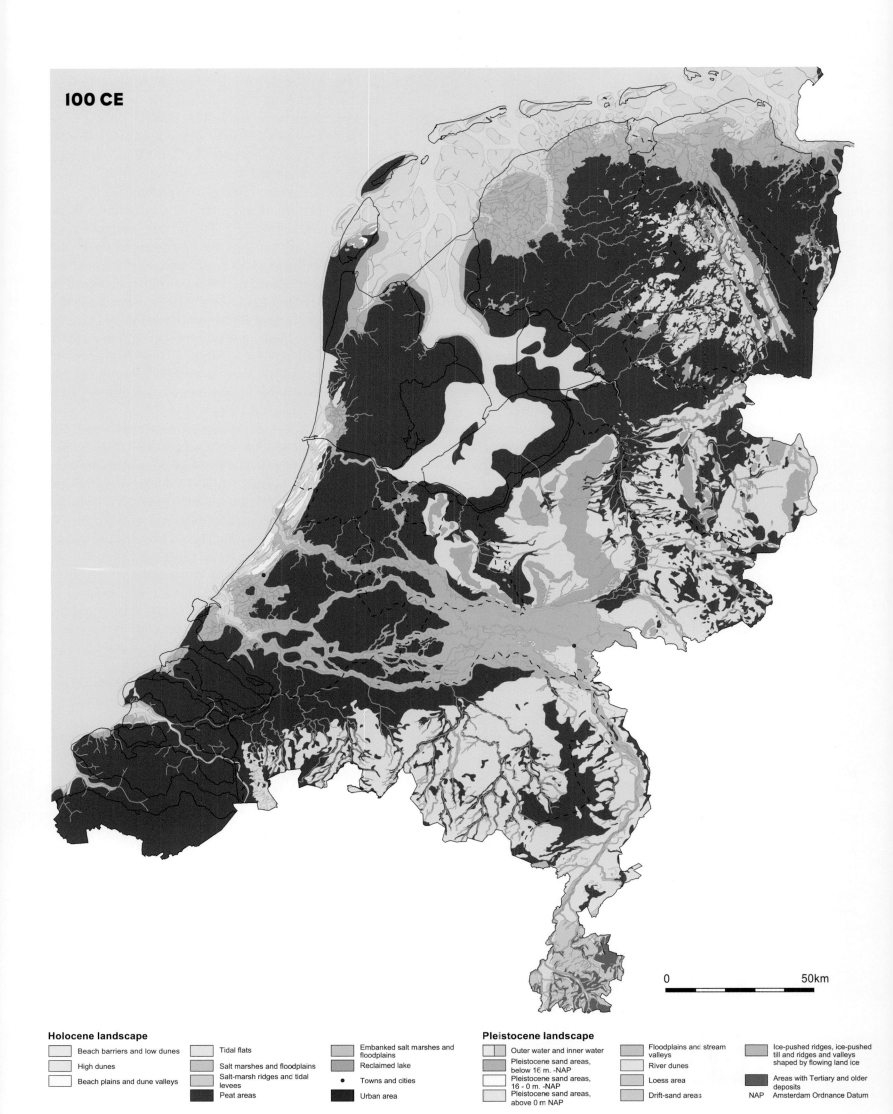

100 CE

Holocene landscape

Beach barriers and low dunes

High dunes

Beach plains and dune valleys

Tidal flats

Salt marshes and floodplains

Salt-marsh ridges and tidal levees

Peat areas

Embanked salt marshes and floodplains

Reclaimed lake

Towns and cities

Urban area

Pleistocene landscape

Outer water and inner water

Pleistocene sand areas, below 16 m. -NAP

Pleistocene sand areas, 16 - 0 m. -NAP

Pleistocene sand areas, above 0 m NAP

Floodplains and stream valleys

River dunes

Loess area

Drift-sand areas

Ice-pushed ridges, ice-pushed till and ridges and valleys shaped by flowing land ice

Areas with Tertiary and older deposits

NAP   Amsterdam Ordnance Datum

0   50km

FIG. 53 In the
Roman period (from
the beginning of the
first century until
c. 400 CE) the major
rivers were used for
the transport of
timber from the
middle and upper
courses of the
Rhine and Moselle.
Raftsmen used
spikes to stay
upright on the log
rafts. This pair was
found in the cabin
of a second-century
flat-bottomed vessel
at De Meern.

They had discovered that dried peat (turf) was an excellent source of fuel and that salt could be extracted from peat that had been flooded by seawater (in a process known as *moernering*). The excavated peat was first dried and then burned, leaving behind the salt.

The landscape changes around the start of the first century CE weren't confined to Zeeland. The landscape also changed in the estuaries of the Meuse, Oude Rijn and Oer-IJ and the coastal region of the northern Netherlands. In the Maasmond area, the increases in water levels that occurred after 250 BCE had caused the salt marshes to become highly silted up, making them suitable for habitation. The peat lakes near Delft had become fully accreted and the Gantel was also largely silted up. New ditches and canals were dug in the highly silted salt marshes of Westland and Delfland in order to maintain drainage.

The formation of new river channels in the central part of the river area meant that from about 250 BCE large volumes of Rhine water could once again start flowing to the Maasmond. The Rhine discharge via the Oude Rijn continued to decline in the Roman period. This drop in discharge brought with it a decline in the quantity of sediment that was transported. But because the currents and wave action continued unabated, the delta at Katwijk, which extended into the sea, began to erode.

The situation around the Oer-IJ also changed dramatically after a beach barrier cut if off from the sea. The former tidal area was permanently habitable during Roman times. It no

longer drained to the sea in a northwesterly direction but to the southeast instead. Following the closure on the seaward side, water drained via the former main channel at Velsen to Amsterdam and from there to the Zuiderzee and the Wadden Sea. Drainage deteriorated in the estuary because the water now had to flow a significantly greater distance. This led to a major expansion of the peatland in the 3rd and 4th centuries CE. The wetter conditions were unfavourable for habitation, which disappeared completely from low-lying areas such as the Uitgeesterbroek and Assendelver polders. Only the more elevated beach barriers continued to be inhabited.

In the northern Netherlands the tidal systems of the Boorne, Hunze and Fivel silted up further, leading to an expansion of the salt marshes. The tidal system of the Middelzee, on the other hand, increased because the Boorne was completely accreted and the Middelzee had taken over all the drainage of the hinterland. The Middelzee also took over part of the catchment area of the Marne in West Friesland. Given its straight course, the connection between the Marne and the Middelzee north of Sneek was probably dug by human hand.

The tidal inlet of the Lauwers between Friesland and Groningen gained steadily in importance as it took over part of the drainage system at Paesens.

The new, highly silted salt-marsh areas of the Boorne, Hunze and Fivel became inhabited in Roman times. The peat margins of these systems were also opened up on a large scale and used for habitation. Here, artificial drainage via ditches and canals had the same result as in Zeeland: the ground subsided and the sea gained access via the excavated infrastructure. At high tide, the land became inundated and a clay layer was deposited over the peat. As in Zeeland, this 'transgressive' development was the result of human intervention.

In the river region, the silting-over of the floodplain peatland continued, a process that was linked to large-scale deforestation in the German hinterland. As a result, the layer of clay overlying the peat increased in thickness and expanded westward.

This silting-over of parts of the peatland and the expansion of the peat lakes meant that the total area of peatland in the Netherlands declined for the first time.

As in the river area, we also see a growing human impact on sedimentation in southern Limburg. The large-scale deforestation and the reclamation of fertile loess soils flushed loess from the slopes and plateaus to the stream valleys and the valley of the Meuse.

The people inhabiting the Netherlands in 100 CE had to contend not only with landscape changes. Caesar's actions in the southern Netherlands were possibly of a genocidal nature. As part of Roman border policy, the Batavians, originally from Hessen, were given a dominant position in the river region in the final decades before the turn of the new millennium. Shortly before the first century CE, Roman troops had made their definitive entry and the southern part of the Netherlands became incorporated into the Roman empire. In the course of the first century CE, the northern imperial border (the *limes*, a modern concept), was located at the Rhine, the northern-most river in the Rhine delta (now the Nederrijn, Kromme Rijn and Oude Rijn). Although the Netherlands occupied only a peripheral position within the Roman empire

FIG. 54 Reconstruction of a
Roman 'villa' from
Rijswijk-De Bult
(Archeon, Alphen
aan den Rijn), a
stone and timber-framed byre
house from the
second century
CE.

and was primarily important as a route connecting the Rhineland and Britannia, its absorption into a state structure triggered various processes of change in the political, social, cultural and economic spheres. These processes, also known as Romanisation, were reflected in the ways in which people organised and engaged with the landscape, especially south of the Rhine. In the zone north of the Rhine, customary styles of habitation and landscape use continued in more or less the same way.

The population grew considerably in Roman times. The number of settlements rose and existing settlements expanded, particularly in the river area and on the southern sandy and loess soils. Choices about where to live were still largely governed by the natural environment. People preferred to build their settlements at dry, elevated locations, close to fields and pastures. There was also a growing diversity and hierarchy within the settlement system. Noviomagus (Nijmegen) and Forum Hadriani (Voorburg) became the first official cities and functioned as administrative centres for the tribal areas of the Batavians and Cananefates respectively. In addition, settlement cores developed in the surrounding countryside and played a key role in trade, artisanal activity and religion. But the vast majority of settlements continued to be centred on agrarian activity .

The rising population and increasing complexity of society boosted the overall demand for products and for a larger product range. This led to greater specialisation and to production aimed at the market place. The fertile loess soils were put into use for the large-scale production of bread wheat and spelt to supply the military and the emerging urban market. This area also saw the appearance of a new type of settlement: the *villa*, or large country estate. People on the sandy soils of Brabant and the river area tended to continue living in byre houses, although these were adapted to Roman fashions. Mixed farming formed the basis of agriculture there, while in the west this was primarily livestock farming.

Roman influence on the fauna was evident first and foremost in cattle and horses, which were much larger thanks to the importation of different breeds and to selective breeding. The Romans also introduced new species: chicken became an integral part of the diet, cats were the new pets and black rats became an enduring part of our fauna. Roman food preferences led to the planting of walnut trees and sweet chestnuts. At around the start of the first century CE the hornbeam became an indigenous species, possibly because of a temporary warming of the climate. This would be the last tree species to reach our country under its own steam.

With these changes, people began to make larger-scale interventions in the landscape and in a way that was more centrally organised. The Romans built an extensive road network for the transportation of soldiers and goods. As key transport and trade routes, the Rhine and Meuse were equipped with harbours, quays and jetties. And in some places, hydraulic works were constructed, such as canals, dykes, culverts and dams. The Romans also built a series of forts along the Rhine, linked by an artificial road. At first, the material they needed for this, timber in particular, was taken from the immediate vicinity, which must have made major inroads on the local tree stock. In the 2nd century the construction timber came from the middle and upper reaches of the Rhine and Moselle,

while most of the stone building material was imported from Germany and France.

By means of a comprehensive system of ditches, the landscape in much of the southern Netherlands was systematically parcelled up for agricultural purposes. But the soil didn't lend itself to intensification in all areas, as evidenced by the reclamations in Zeeland, where ground subsidence and saturation occurred. Sandy soils may have become degraded as a result of more intensive farming practices. A clue here is the introduction of sunken byres, which made it possible to capture animal manure.

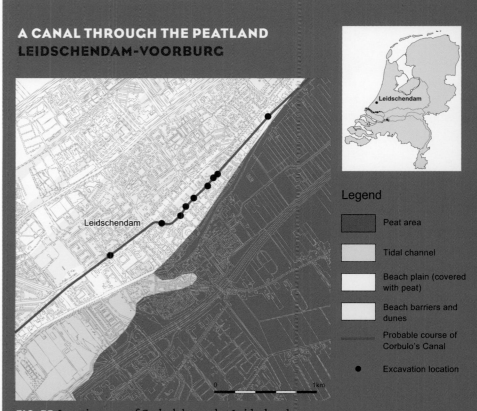

A CANAL THROUGH THE PEATLAND
LEIDSCHENDAM-VOORBURG

Leidschendam

Leidschendam

Legend

Peat area

Tidal channel

Beach plain (covered with peat)

Beach barriers and dunes

Probable course of Corbulo's Canal

● Excavation location

0   1km

FIG. 55 Location map of Corbulo's canal at Leidschendam.

From the Roman period, people became increasingly adept at managing nature to their own advantage. A good example is Corbulo's canal. In 50 CE, to keep his soldiers occupied, the Roman army commander Corbulo ordered the construction of a canal linking the Meuse and Rhine. This created a safe route for military transport and trade, avoiding the perilous North Sea.

The canal was dug roughly between present-day Leiden and Naaldwijk. Its location was very much governed by the landscape: it ran through the peatland immediately behind the eastern-most beach barrier, where the peat could be easily excavated. A recent discovery shows that the canal cut across the beach barrier in Leidschendam. West of that point, it then continued on towards the Meuse, for reasons that are not yet clear.

At weak locations, oak posts – probably from trees on the nearby beach barrier – were used to reinforce the sides of the canal.

The canal linked two natural tidal creeks: one running southwards from the Rhine and a side-creek of the Gantel. In other words, the canal didn't cover the full distance: the soldiers only built it where there were no natural water courses.

In the first century CE, a city and harbour sprang up on the canal. It was called Forum Hadriani (Voorburg) and was the capital of the tribal area of the Cananefates. The harbour played a key role in supplying the forts and settlements along the coast. However, its significance could not prevent the canal from largely silting up a short time later.

800 CE
RETURN OF THE SEA

The map of 800 CE shows how the sea had tightened its grip on the land: Zeeland had drowned and, in the northwest, there was now a much wider connection between the Wadden Sea and the IJsselmeer region and the peat lakes had expanded enormously. The peatlands in the river area had disappeared beneath a clay cover. In the meantime, the departure of the Romans triggered a significant drop in population after the third century. Human impact on the landscape also declined, although this was just temporary. When the population rose again several centuries later, the inhabitants once again began to fell trees on a vast scale and to build water-management systems.

In 800 CE the sea had returned in full force. In the southwestern Netherlands, human intervention had caused a large tidal area to evolve, which reached its maximum extent in about 800 CE. Large-scale Roman time reclamations, involving the draining of peatland, had caused significant subsidence. This process was exacerbated by the fact that the sea could penetrate deep into the peatland via drainage channels and ditches. Ultimately, these reclamations had catastrophic consequences: the peat flooded, there was a big increase in tidal volume, the tidal channels widened and eventually the sea swept away large chunks of peat along the tidal inlets. The eroded peat wasn't deposited elsewhere: unlike sand, eroded peat is lost to the sediment balance because it consists largely of water and the remaining organic part decomposes.

The process of subsidence, increasing tidal volume, expanded tidal channels and peat erosion was self-reinforcing, causing the Zeeland peat to disappear at a great rate. In about 350 CE the Roman peat reclamation areas were completely inundated, making habitation and further peat extraction impossible. Later, the non-reclaimed peatlands were also sub-

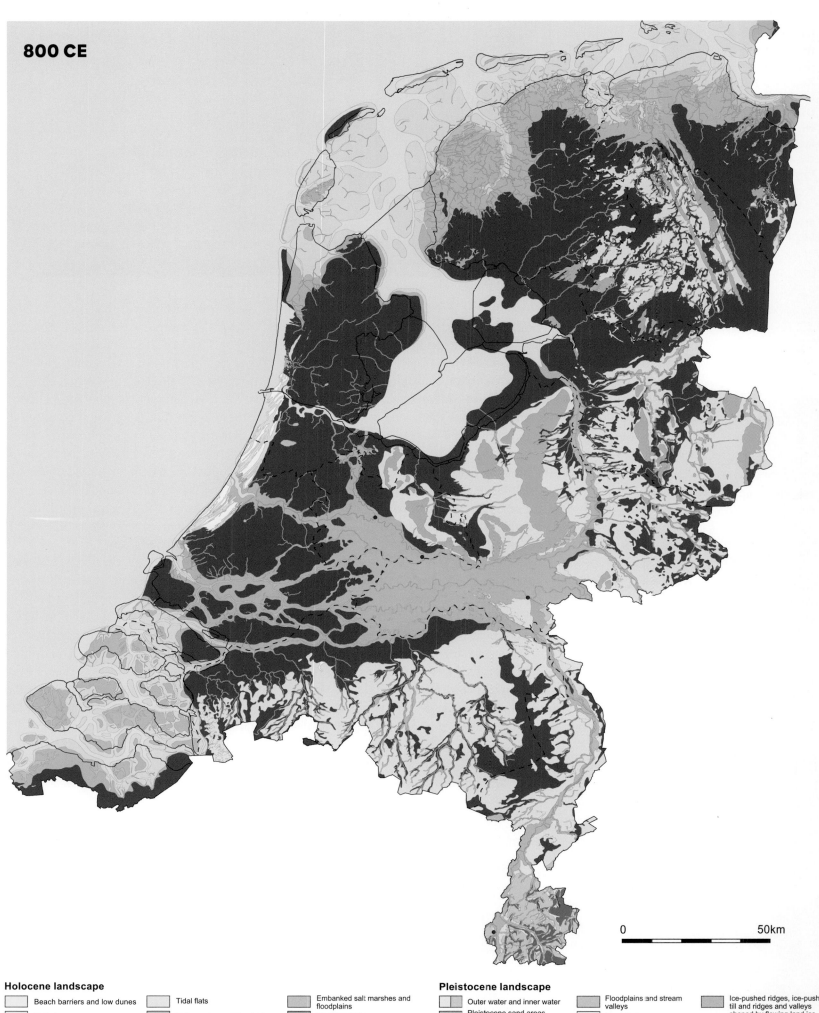

800 CE

FIG. 57 The early medieval town of Dorestad was partly an agrarian settlement and partly a trading post on the Rhine. Because the river, the present-day Kromme Rijn, kept changing course, the inhabitants were obliged to keep extending the wooden jetties further out into the river. The jetty remains were excavated at Wijk bij Duurstede.

merged. On the other hand, after centuries of sedimentation the areas that were the first to be flooded began to silt up again to salt-marsh level. The mouth of the Scheldt had become an estuary at the site of present-day Oosterschelde.

When the southwestern Dutch tidal area was being formed, waterways were also opening up to the Rhine-Meuse delta. Because not all of the river water now flowed to the sea via the Maasmond, the peat could expand once more along that river mouth.

The Oude Rijn had become a minor branch of the Rhine that transported barely any sediment. Erosion then caused the river delta that had formed in the sea at Katwijk to slowly but surely disappear. Wave action carried some of the sand from the delta to the former mouth of the Rhine at Katwijk, which then began to silt up.

The closing of the Oer-IJ estuary in the late iron Age also caused the inlet to silt up. The coastline of the western Netherlands now ran in a straight line. Poor drainage in the former tidal zone of the Oer-IJ led to a considerable increase in peat formation in this area. The former main channel of the Oer-IJ between Velsen and Amsterdam continued to widen as the peat along the edges was eroded, thereby creating the IJmeer. The deep former main channel filled with peat detritus as a consequence of the peat erosion.

In the first millennium, the Zuiderzee had become a large inland sea as the Flevo lakes merged to become one. This process was the result of tides in the estuary and wave action, which caused the edges of the peat areas to crumble away. The larger the expanse of water, the greater the wave action.

Texel was cut off from the mainland in about 800 CE when a tidal area formed there. Before that, Texel had always been attached to the rest of North Holland by a beach barrier, with a peat area behind it. However, the sea had washed this away and a tidal zone developed between the island and the shore.

The tidal area between Texel and Friesland, which evolved after about 500 BCE, had expanded. This was mainly due to the drowning of the peatland in northern North Holland, in which humans may also have played a part.

There were also new sea incursions in the northern Netherlands. These were largely the result of large-scale early-medieval reclamations in the peat hinterland, including along the shores of the Middelzee. Flooding caused a salt-marsh clay layer to form there, creating a clay-on-peat area. Because the seawater would force its way through the low-lying hinterland during storms, the associated channel system, the Middelzee, was greatly expanded. A similar process occurred in the peat margins between Friesland and Groningen, where a new tidal system, the Lauwerszee, developed.

In the meantime, the northern tidal areas almost completely silted up, causing the marshlands to expand significantly at the expense of the mudflats that were submerged daily by the tide. These marshlands, where people could safely live on terps, became favoured locations for habitation.

In the river area the covering clay layer expanded still further westward, to beyond the line between Utrecht and Gorinchem. Various rivers had changed course in the Roman period: the Rhine (at Wijk bij Duurstede), the lower reaches of the Waal (at Zaltbommel) and the Meuse (from Den Bosch onward). This made the downstream part of the delta much wetter; the new rivers flooded frequently because they had not yet had an opportunity to form high banks. The Gelderse IJssel probably appeared in about 800 CE. Following a breach, possibly between Zutphen and Deventer, the Rhine was now connected to the large stream that flowed through the later IJssel valley.

FIG. 58 Reconstruction of a farmhouse from the northern Netherlands terp region (Yeb Hettinga Museum, Firdgum). The lack of timber for building in the treeless salt-marsh region necessitated the widespread use of salt-marsh sods in house construction.

Soil erosion continued in southern Limburg until the end of the Roman era. When the Romans left, much of the agricultural land was abandoned, and there was a decline in erosion and the associated loess sedimentation.

The third century CE brought an end to the heyday of the Roman empire. The consequences for habitation were at least as drastic as the landscape changes that were happening at that time. For two centuries the region was marked by a state of decline. Population numbers fell dramatically in the former Roman territories, while urban agglomerations shrank or even disappeared altogether. All the villas – the large farming estates in the Roman countryside – ceased to exist and many other settlements were abandoned as well. Population numbers also fell north of the *limes* and some regions, such as the northern coastal area, appear to have been abandoned altogether for a time. In the settlement cores that survived, agrarian activity probably dropped to the level it had had before the Roman occupation. In the south especially, uncultivated land appears to have become forested once again. But for some remarkable animals it was too late: aurochs and large birds like the black vulture and great auk had vanished from our country.

The demographic decline didn't stop until the sixth century, when our regions gradually came under the sway of the Merovingian kings, whose heartland lay in France. In subsequent centuries they and their successors, the Carolingians, were able to significantly expand their power base and their territory, at the expense of the Frisians and Saxons. At the time of Charlemagne's reign (c. 800 CE), our entire country was part of the Carolingian empire.

The Merovingian period was characterised by a rise in the number of settlements. There were clear concentrations near waterways and in areas with good agricultural potential. However, not all parts of the country were habitable at this time. Although some of the marginal coastal peat zone was reclaimed, a vast area of peat still remained. Living there was impossible, with the exception of a few places in Kennemerland and western Friesland and on the marine deposits on the Zeeland coast.

In areas that fell under Frankish rule, a process we call 'Frankicisation' took place. This involved the introduction of new political, socioeconomic and religious systems, including Christianity, in the conquered territories. Frankicisation also entailed a new ordering of the landscape. In the Carolingian period, land lay at the heart of power and prestige. The available lands were redistributed among the elite and ecclesiastical institutions, and settlement cores acquired a permanent place. Large estates were organised into manors, large agrarian complexes that were managed from central courts. Once again, huge numbers of trees were felled in order to clear the land for agriculture. Another outcome of Frankicisation was the introduction of a wide range of new fruit species, such as medlars, pears, quinces and mulberries.

The population in the inhabited areas grew during the Carolingian period, and uninhabited regions were once again put into use or colonised. This latter development is especially evident in North Holland, where the first large-scale peat reclamations began. The newly silted-up parts of the Zeeland islands were also resettled in this period. As well as agrarian

settlements, suburban centres developed, such as Deventer, Dorestad, Medemblik, Maastricht, Nijmegen and Utrecht. The latter three had already been inhabited in Roman times and were probably never completely depopulated. Port facilities were constructed at trading centres during this period, as well as some incidental and very modest water-engineering works.

Iron production was a major activity throughout this time. In the late Roman period this mainly occurred in parts of the eastern Netherlands, and later also in the Veluwe and the Gelderse Vallei. In some regions, Carolingian-period mining led to large-scale deforestation to produce the immense quantities of charcoal needed for the smelters.

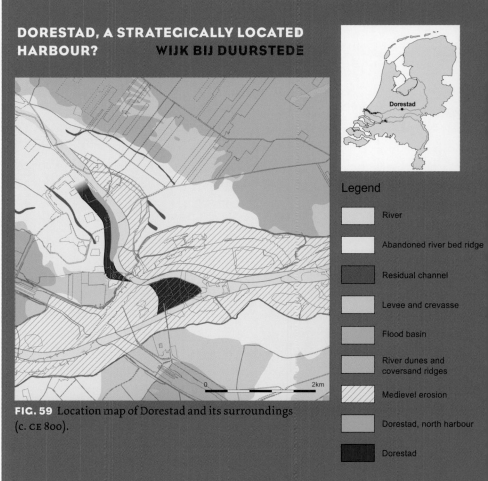

DORESTAD, A STRATEGICALLY LOCATED HARBOUR?   WIJK BIJ DUURSTEDE

Legend

- River
- Abandoned river bed ridge
- Residual channel
- Levee and crevasse
- Flood basin
- River dunes and coversand ridges
- Medievel erosion
- Dorestad, north harbour
- Dorestad

FIG. 59 Location map of Dorestad and its surroundings (c. CE 800).

In about 800 Dorestad (now Wijk bij Duurstede) was one of the major trading centres of northwestern Europe. The settlement was strategically located at the fork of the Rhine and the Lek, an arterial route linking Scandinavia and the British Isles with the German Rhine area and the Meuse valley.

Goods from all corners of the globe came into Dorestad harbour: honey, salt, wine, basalt quern-stones, pottery, glassware, grindstones, weapons, jewellery, furs and hunting dogs, as well as raw materials such as timber, iron ore, amber and glass ingots. These and locally available raw materials such as bone and antler were worked by craftspeople into end products, including bone combs, glass and amber beads and iron tools. All these

products could be bought in Dorestad, along with regional products such as cereals, vegetables and fruit, meat and fish. Some of the goods were sold to the local population in the Kromme Rijn region, while others were shipped to customers in sometimes distant parts.

Dorestad's favourable location as a transport hub also had a downside. In the years 834 to 863 the town became the target of plundering bands of Vikings. It lay within easy reach of their boats and was repeatedly devastated by Viking raids. The landscape dynamics were just as disastrous: the river – the lifeblood of economic activity – silted up during this period, rendering the harbour virtually inaccessible. In about 875 the trading heart stopped beating altogether.

1250 CE

DYKING OF RIVERS AND SALT MARSHES

FIC. 60 Medieval dyke along the Eem river at Eembrugge, with a kolk lake (*wiel*) – the remains of a dyke breach – on the right.

In the high and late Middle Ages, humans became the dominant factor in shaping the landscape of the Netherlands. Peatland was opened up on a vast scale and dams and embankments were built on the high salt marshes and along rivers. Large tracts of land on the higher sandy soils and the loess soils of Limburg were also put into cultivation. It was in this period that the foundations were laid for the man-made landscape that we know today. In a complex interplay with nature, the people of this period set in motion major landscape changes, whose consequences are visible to this day.

After 800 CE, the substantial supply of marine sediment from the sea caused the salt marshes along the entire coastline to silt up and the tidal channels became partially closed once again. As a result of this accretion process, salt-marsh areas outside the northern Netherlands also became habitable in subsequent centuries. In the tenth century, parts of the newly formed salt marshes fell permanently dry due to a lowering of storm flood levels. Settlements appeared on the elevated salt-marsh deposits.

In the eleventh century, the inhabitants of Zeeuws-Vlaanderen began embanking parts of the salt marshes. By the twelfth and thirteenth centuries, this had developed into a major, systematic undertaking in salt-marsh areas along the entire Dutch coast and in the (former) peatlands. All the major rivers in the river area were embanked, while smaller rivers, such as the Kromme Rijn, Oude Rijn and Linge, were dammed. By about 1250 most of the coastal and river region was dyked. The West Frisian circular dyke (*West-Friese omringdijk*) must have been fully closed at about that time and virtually the entire salt-marsh region at the mouths of the rivers Boorne, Pae-

1250 CE

Holocene landscape

- Beach barriers and low dunes
- High dunes
- Beach plains and dune valleys
- Tidal flats
- Salt marshes and floodplains
- Salt-marsh ridges and tidal levees
- Peat areas
- Embanked salt marshes and floodplains
- Reclaimed lake
- Towns and cities
- Urban area

Pleistocene landscape

- Outer water and inner water
- Pleistocene sand areas, below 16 m. -NAP
- Pleistocene sand areas, 16 - 0 m. -NAP
- Pleistocene sand areas, above 0 m NAP
- Floodplains and stream valleys
- River dunes
- Loess area
- Drift-sand areas
- Ice-pushed ridges, ice-pushed till and ridges and valleys shaped by flowing land ice
- Areas with Tertiary and older deposits
- NAP   Amsterdam Ordnance Datum

0   50km

FIG. 61 Shortly after 1270 CE a dam was built on the Rotte river; the small polder behind it became the nucleus of the later city of Rotterdam. The dam featured large wooden sluices that drained surplus water into the Meuse and prevented water from that river from flowing in.

need to produce more food for the growing population. This exploitation of the peatland created the peat-pasture landscape that is so characteristic of the low Netherlands. The cultivation of the expansive peatlands in the north and west of the country was undertaken by groups of people. Because farmers were more or less free to cultivate the land as they saw fit, these early peat reclamations initially tended to display an irregular parcelling pattern. Farmsteads shifted frequently because the most recently reclaimed tracts were more suitable for cultivation.

Over time, reclamation became more systematic and was increasingly driven by the authorities. This produced the parcelling system that was so characteristic of the peat areas: parallel strips running perpendicular to elongated settlements along roads or dykes. The large peat area held by the counts of Holland and bishops of Utrecht was tackled in a highly orderly fashion. The manner of reclamation was the same as elsewhere in the country, except that the measurements were very strict. The parcel length (*opstrek*) was set at 1250 to 1350 metres. The area roughly between Amsterdam, Rotterdam and Utrecht was therefore divided up very regularly into long strips within rectangular cultivation blocks. Settlements in this region were not relocated. The sale of peatland by the counts and bishops to land reclaimers – via contracts known as *copen* – is reflected in settlement names ending in -cope, such as Boskoop and Nieuwkoop.

In the north, the initiative to open up new land for cultivation came mainly from independent farmers, who later donated their properties to the monasteries. In the IJssel valley, the duke of Gelre was the primary instigator, although not until the fourteenth century.

The large-scale reclamation of the peatlands had major consequences for the landscape. Drainage caused the peat to subside, making the peatland more vulnerable to the sea. At the same time, the dykes increased in length, causing the sea and river water to build up as it could no longer overflow onto the hinterland.

Large-scale peat erosion occurred on the shores of the Zuiderzee, which had almost reached its maximum size by about 1250. The peat area below Wieringen was lost, partly as a result of cultivation, and became part of the Zuiderzee. Natural land expansion in the Zuiderzee occurred only at Kampen, where the river IJssel extended its delta – Kampereiland.

In Friesland and North and South Holland, lakes appeared in the heart of the peatland. They were the result of the peat reclamations and the limited opportunities for drainage. Peat may also have been dug on a limited scale along existing water courses, such as the Purmer, Schermer and Wormer, for use as fuel. In high winds, the water in these peat lakes battered against the surrounding peat, causing it to crumble still further. In the late Middle Ages and early modern era, this gave rise to large lakes which continued to expand over time.

In Zeeland and on the islands of South Holland, we see a

sens, Hunze and Fivel was embanked. In the late Middle Ages, the Middelzee silted up rapidly because the dykes had drastically reduced the size of this storage area. They had also caused a sharp drop in tidal currents, whereby the channel silted up; it too was embanked in subsequent centuries. The Lauwers tidal system was the only one in the northern Netherlands that expanded, partly because the Lauwers had taken over the drainage function from the Hunze. The Reitdiep, linking the Hunze to the Lauwers, had evolved at Zoutkamp. Given the 90-degree bend between the two systems, it is quite possible that this connection was dug by human hand. Now that it was connected to the Lauwers, the mouth of the Hunze rapidly closed over and the newly formed salt marshes were embanked.

From the tenth century onward, the peatlands behind the coast were systematically taken in hand in response to the

FIG. 62 De Armenhoef in Best is probably the oldest farmhouse in western Europe. Tree-ring dating has revealed that the house's timber frame dates from 1263. The walls, window frames and roof date to the nineteenth and twentieth centuries.

phenomenon quite unlike what occurred in the western and northern Netherlands. Much of the peatland had disappeared or been overlain with clay and sand. The underlying peat obtained a high salt content due to inundation by the sea, making it ideal for salt extraction (*moernering*, see text accompanying the map for 100 CE). The emerging Flemish cities were driving the demand for salt. The extraction pits can still be seen in the landscape today.

It was not only the low Netherlands that underwent major changes in this period. Stream valleys on the sandy soils were increasingly equipped with watermills and much of the uninterrupted tracts of forest disappeared. The wood was turned into charcoal, some of which was used for the production of iron. From the eighth to the eleventh centuries, many tonnes of iron were produced, particularly on the Veluwe and in Montferland.

The forest on the sandy soils was replaced by a more open woodland vegetation with expanding heathland and occasional fields. Older, more scattered fields or *kampen* were joined together to form extensive agricultural complexes. These *essen*, also called *engen*, *enken* or simply *akkers* (fields), were intensively cultivated. The associated farmhouses were often grouped in a hamlet at the centre of the *es* or were located in a more scattered fashion along the fringes of the field complex. The increase in field area was accompanied by improvements to the plough and the development of the horse collar (*haam*), a wooden yoke that – unlike earlier horse gear – didn't interfere with the horse's breathing.

This more intensive exploitation of the sandy areas also had its downsides. In some places the sand began to drift, a problem that would only worsen in the late Middle Ages and early modern era, mainly as a result of human influence on the vegetation. These sand drifts would sometimes pose a threat to settlements and to farmland.

The greater population density and increased area of cultivated land also led to the disappearance of the last large wild animals, such as elk, red deer and bears. Sea eagles died out too. On the other hand, this period saw the arrival of rabbits, imported from southwestern Europe in the thirteenth century for hunting.

The foundations of today's settlement pattern were laid in this period. Habitation in the countryside still largely took the form of scattered farmhouses and small hamlets. Up until the High Middle Ages, these settlements shifted regularly. This phenomenon came to an end in the thirteenth century: people stayed living in the same location, in places where habitation still continues today. A process of village formation began, in most cases prompted by the founding of a church. True villages – larger settlements that included artisans – were initially few in number: it wasn't until the fourteenth and fifteenth centuries that most villages began to make their appearance.

Trading settlements developed along the rivers. Some grew rapidly, especially from the twelfth century onwards, and evolved into cities such as Tiel, Deventer and Medemblik. In other places – on the coast and inland – cities didn't start to develop until the thirteenth century. Haarlem, Dordrecht, Den Bosch and Roermond are just a few examples. Settlements arose around dams in peat rivers – essential for draining the newly reclaimed peatland of this period – and later grew into cities. Examples include the dam on the rivers Amstel (Amsterdam), Rotte (Rotterdam), Zaan (Zaandam) or Schie (Schiedam). Numerous castles were also built at this time, for example in the Kromme Rijn region. They were commissioned by the nobility, who had acquired fortunes and prestige by opening up new land or by some other means. For the first time since the Roman period, monumental buildings were once again erected in the Dutch landscape alongside the modest byre houses: in addition to the castles just mentioned, there were many churches and monasteries (especially from the twelfth century onwards). This time they occurred not only south of the big rivers, but throughout the Netherlands.

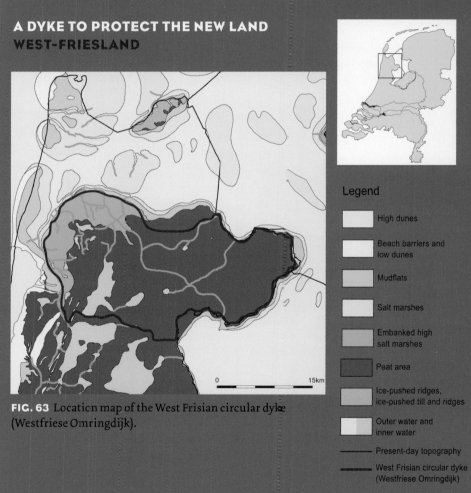

A DYKE TO PROTECT THE NEW LAND
WEST-FRIESLAND

Legend

High dunes

Beach barriers and low dunes

Mudflats

Salt marshes

Embanked high salt marshes

Peat area

Ice-pushed ridges, ice-pushed till and ridges

Outer water and inner water

Present-day topography

West Frisian circular dyke (Westfriese Omringdijk)

FIG. 63 Location map of the West Frisian circular dyke (Westfriese Omringdijk).

When West Friesland was colonised in the Middle Ages, people encountered a large peat area that rose high above sea level. Reclamation and drainage made it possible to grow rye there, but this cultivation led to oxidation and subsidence. In a fairly short time, the soil level dropped to below the highwater mark and dykes became necessary. As preparation for dyke building, a layer of clay may have been dumped on top of the peat to 'preload' it and prevent the dyke from being washed away with the underlying peat layer. Salt-marsh sods were placed on top of the clay layer, thereby extending both the height and width of the dyke at different stages. These extensions were necessary because the dyke body sank into the peat under its own weight and the peat area itself also subsided further.

The origins of the West-Friese circular dyke are shrouded in mystery, however. The dyke is first mentioned in historical sources in the final decade of the 13th century. It is likely, however, that the initial construction goes back to the second half of the 12th century and first half of the 13th century. Its construction required a major investment of collective labour on the part of local farming communities. How and under what authority they were united is unclear. After the conquest of West Friesland by count Floris V, a bailiff – a predecessor of the dyke reeve – was appointed to supervise the building and maintenance of the dyke.

PEOPLE SHAPE THE LANDSCAPE

FIG. 64 Parcelled land at Kockengen (Utrecht province). These characteristic elongated plots date back to the medieval reclamation of the area.

The large-scale human interventions in the High and Late Middle Ages brought about major changes, both along the coast and inland. The coastal peat, which had originally been several metres higher, had sunk to sea level as a result of drainage of the peatland. In addition, the continuous embankments meant that the sea and river water had nowhere to escape to when water levels were high. The combination of soil subsidence in the embanked peat areas and water building up against the sea and river dykes sometimes led to catastrophic dyke breaches and flooding. On occasions, land was lost to the water. On the other hand, much of the drowned land had been reclaimed from the sea before 1500, thanks in part to new embanking techniques and the use of drainage windmills.

Between 1250 and 1500 much of the coastal and river region had become one vast area fully enclosed by dykes and with its own water management system. Surplus water was drained artificially. Initially, it was enough to simply open a sluice if water levels outside the dyke were low. But as the ground surface inside the dykes continued to subside, the point at which drainage could be regulated in this way lay increasingly further afield. Long drainage ditches had to be dug. In places where natural drainage was no longer possible, additional mechanisms such as handmills and horse-driven mills were pressed into service to drain the water. Windmills were first used for this purpose shortly after 1400.

The continuous line of embankments, which was largely complete before 1250, had a considerable impact on the landscape in unembanked areas. Although the dykes prevented the hinterland from being flooded at times of high river discharge or spring tide and storm, the sea and river water banked up against the dykes because it had nowhere else to go. The areas outside the dykes also rose higher through silting, while the peat behind the dykes continued to subside. In West-

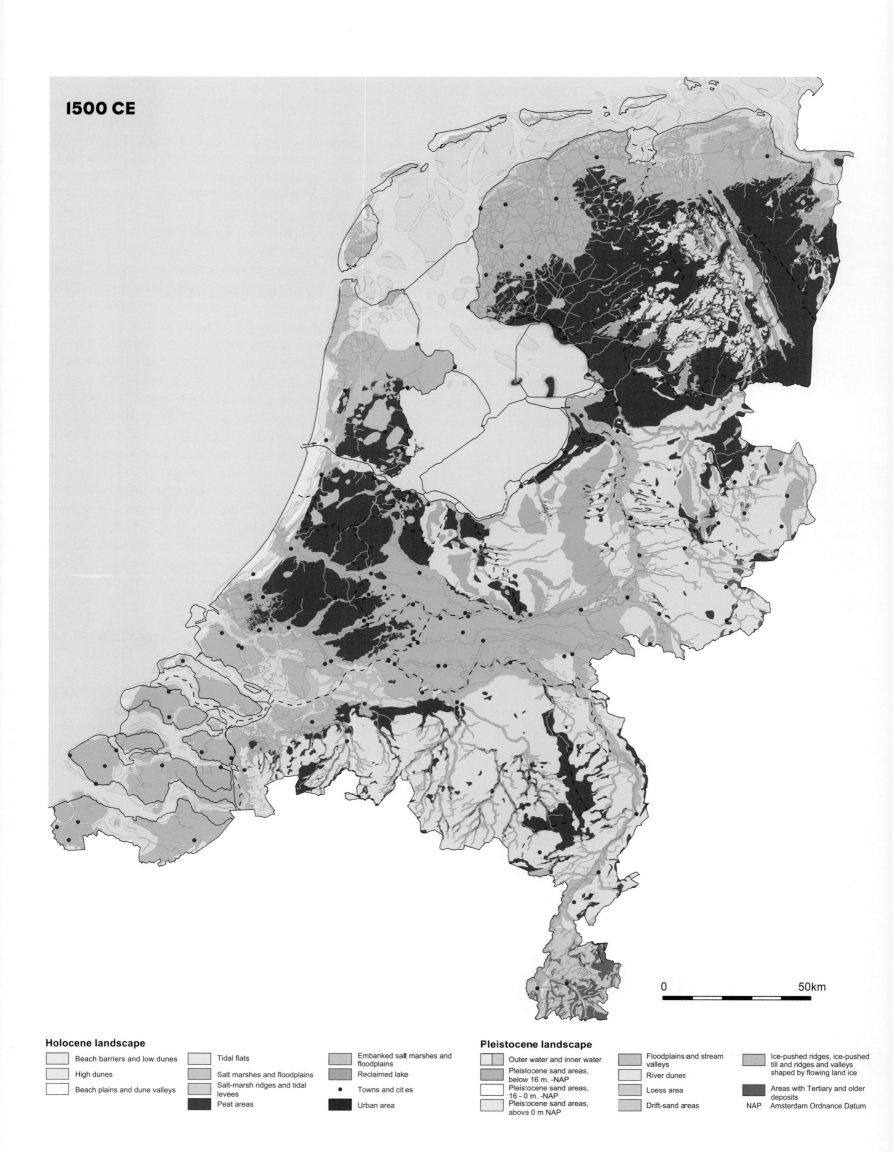

1500 CE

Holocene landscape

- Beach barriers and low dunes
- High dunes
- Beach plains and dune valleys
- Tidal flats
- Salt marshes and floodplains
- Salt-marsh ridges and tidal levees
- Peat areas
- Embanked salt marshes and floodplains
- Reclaimed lake
- Towns and cities
- Urban area

Pleistocene landscape

- Outer water and inner water
- Pleistocene sand areas, below 16 m. -NAP
- Pleistocene sand areas, 16 - 0 m. -NAP
- Pleistocene sand areas, above 0 m NAP
- Floodplains and stream valleys
- River dunes
- Loess area
- Drift-sand areas
- Ice-pushed ridges, ice-pushed till and ridges and valleys shaped by flowing land ice
- Areas with Tertiary and older deposits
- NAP   Amsterdam Ordnance Datum

0   50km

FIG. 65 As early as the Iron Age and Roman period, people extracted salt by heating sea water on turf-fuelled fires. Already then, peat-digging for salt production probably had disastrous consequences for the landscape. This was certainly the case with the late medieval and early modern practice of *moernering*, whereby salt was extracted by burning peat that was saturated in sea water.

Friesland, for example, which had been enclosed by a dyke since 1250, the fairly thin peat layer had entirely disappeared by 1500. In the coastal area of Zeeland and the islands of South Holland, the practice of digging peat for salt extraction also continued in this period, causing a further lowering of the ground level.

FIG. 66 The Netherlands was rapidly urbanised during the late Middle Ages, a process accompanied by a host of new house types and ways of living. One famous dwelling is Begijnhof 34 in Amsterdam, which dates from about 1530. It is a very rare example of an urban house with a wooden façade.

The lowering of the ground level following peat reclamation and turf extraction led to a reduction in the total peat area because the peat lakes kept increasing in size. Once the peat had disappeared, many of the peat areas were lost through flooding. Attempts were made, with varying success, to reclaim the land once more. Peat and salt-marsh areas where the ground level was above mean high water level in times of flood dried out again quickly and could therefore be embanked again fairly easily. However, peat areas and salt marshes whose surface was below mean high water level during flood were much harder to drain again. This was the case with the Verdronken Land van Zuid-Beveland, De Braakman in Zeeuws-Vlaanderen, the Verdronken Land van Saeftinghe, the Zuidhollandsche Waard and the Dollard. Their location below mean high water level meant that water flowed in and out of the drowned area at each tide. The larger tidal channels that resulted could not be closed immediately with the resources available at the time. Dykes and dams couldn't be built there until the land silted up again sufficiently over time and the channels accreted.

The Biesbosch is one area where storm surges were able to inflict major, long-term damage because of the changes that people made to the landscape. It was once a vast polder landscape: the Groote (or Zuidhollandsche) Waard. The dykes there had already become weakened in the early 15th century as a result of earlier storm surges and neglect. However, they were so severely affected during the famous St Elizabeth Flood of 1421 that the area had to be abandoned in the wake of a subsequent heavy storm surge in 1424. Because both the sea and river dykes along the Merwede were breached, the river now flowed across the area, creating a large freshwater tidal zone. Silting caused parts of this zone to become dry again later and the Biesbosch acquired its present-day form.

Another severe storm surge was the St Felix Flood of 1530, which led to the permanent loss of large parts of Zuid-Beveland (Drowned Land of South Beveland) and the creation of the Westerschelde. This caused a new watershed between Bath and Woensdrecht and the Oosterschelde to silt up rapidly because of the decrease of the tidal currents. The Westerschelde now became the main connection to the sea. The impact of human activity is most evident today in the Verdronken Land van Saeftinghe. This former peat region, which lay below mean high water level, was deliberately flooded during the

Spanish siege of Antwerp in 1583. The Sea Beggars of Holland (*Hollandse Watergeuzen*) pierced the dykes and the sea did the rest. The area looks like a natural salt marsh area to this day.

During this period the sea also made several major incursions into the northern Netherlands, leading to considerable loss of land. Between the 13th and 16th centuries, major sea floods occurred in the Dollard region of northeastern Groningen. The last one was the Cosmas and Damianus Flood in 1509, which gave the Dollard its maximum size. The floods were the result of large-scale peat reclamation in an area where the Pleistocene valleys were filled with a metres-thick layer of peat. The substantial thickness of the peat meant an acceleration of the ground subsidence, making the area extra-vulnerable in times of flooding. During the last major flood, large peat islands that had been torn loose from the subsoil floated away. A peat island, complete with its cargo of livestock, was reported as floating adrift by Ubbo Emmius, rector magnificus of the Academy in Groningen (the predecessor of the University of Groningen), before going aground some distance away. This event sparked a lawsuit as to who owned the floating land.

In the more elevated sandy regions of the Netherlands, virtually all the remaining closed forest disappeared in this period to make way for fields and intensively grazed, open woodlands with much heathland. In the late Middle Ages and early modern era, this led to soil depletion on the field complexes, some of which had been created before 1250. This was counteracted by the collection of litter from elsewhere in the landscape or by digging sods and spreading them, mixed with animal manure, over the fields as compost. As a result, the fields eventually rose above the surrounding landscape. The more intensive sod-stripping and grazing of the open heath caused an increase in sand drifts. As with the situation regarding flood protection, people were not entirely defenceless in the face of these changing conditions. Action was undertaken early on to curb further deforestation and sand drifts through such measures as strict regulation of the use of the heathland. In the Veluwe, a sand reeve (*zandgraaf*) was even appointed to coordinate the battle against the drifting sands.

Archaeological research in the Kempen region illustrates just how much this period influenced the present-day layout of sandy soil areas. Before 1250, the fields and country roads were still characterised by curving contours that had evolved in organic fashion. These are only visible in broad outline in the present-day landscape. By about 1500, the landscape more closely resembled the much more angular and rational pattern of parcels and roads that we have today.

With the exception of the steepest slopes, all of southern Limburg had been opened up for agriculture by about 1500. Once again, this led to a significant increase in soil erosion, with large quantities of loess soils being deposited in the stream valleys.

The settlement pattern acquired a more definitive form during this time. The number of towns and villages rose sharply in the 14th and 15th centuries. By about 1500, the vast majority of today's towns and villages were already in existence. Flemish cities had reached their peak in the 14th century but thereafter stagnated or declined. The towns and cities of Holland and Zeeland, however, grew steadily. By the begin-

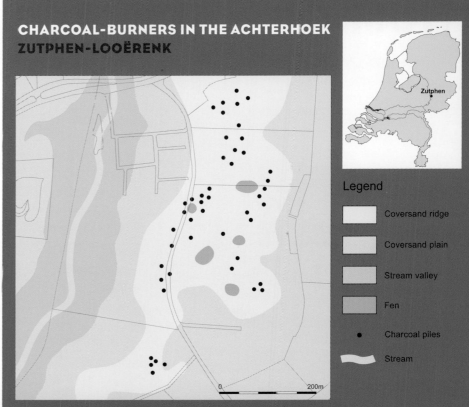

CHARCOAL-BURNERS IN THE ACHTERHOEK
ZUTPHEN-LOOËRENK

Legend

▢ Coversand ridge
▢ Coversand plain
▢ Stream valley
▢ Fen
● Charcoal piles
〰 Stream

FIG. 67 Location map of charcoal pits (eighth and ninth centuries) on the Looërenk at Zutphen.

The Looërenk, a high sandy ridge at Zutphen on the boundary between the IJssel valley and the coversand landscape of the Achterhoek, was the scene of major activity in the eighth or ninth century. The mature oak forest that once stood there had been completely felled. For a time, the deforested terrain was used for grazing before eventually being converted into arable land.

The wood was used to make charcoal. Professional charcoal-burners would stack the wood in pits (*meilerkuilen*), cover it with turf, and then set it alight, leaving it to smoulder until it turned to charcoal. The charcoal from the Looërenk was probably used to smelt iron from ore extracted from the marshy depressions in the vicinity.

The oldest indications of charcoal production date to about the beginning of the first century. In the Veluwe, Montferland and the western Achterhoek, this craft developed into a major industry during the early Middle Ages. The exceptional scale of charcoal production at Zutphen, where there were hundreds of these charcoal piles, and at Anloo in Drenthe, may indicate the direct involvement of the new Frankish overlords. In the later Middle Ages aboveground charcoal piles increasingly replaced the use of pits. Charcoal-burners were active until the twentieth century in forested parts of the Netherlands.

ning of the 15th century, a significant proportion of the population there was already living in cities. This situation is particularly unusual if we compare it with what was happening elsewhere in Western Europe. Although precise population figures are missing, it seems that the 14th and first half of the 15th century saw an end to the strong population growth of the previous centuries virtually throughout Western Europe. Population numbers fell almost everywhere and thousands of settlements and vast tracts of cultivated land were abandoned.

1850 CE
HUMAN 'NATURE'

By 1850 the Netherlands had more or less acquired its present form. Today we regard many of the landscapes of that time as 'natural', but they were almost all the result of far-reaching human intervention on an accelerating scale. The upsurge in prosperity that the Netherlands had enjoyed after 1500 was clearly reflected in the landscape.

In about 1500 humans had already become the main 'geological factor'. Almost everywhere, the shape of the landscape was partly due to human intervention – whether intentional or otherwise. The primeval forest had virtually disappeared and the consequences of peat reclamation and water control were clearly visible.

The map of 1850 shows that the human hold on the landscape had tightened significantly since the Middle Ages. Between 1500 and 1850 major land reclamation occurred along the entire coast – on the islands of Zeeland and South Holland and in northern North Holland, Friesland (Het Bildt) and Groningen (around the Dollard). This was possible thanks to the natural silting-up of areas outside the dykes. Once these areas had silted up high enough, they were embanked and used as grazing land or for crop cultivation.

Running counter to this silting-up of the salt-marshes was the erosion of the coast itself. The western Netherlands coastline had been eroding since Roman times and had shifted several kilometres eastwards. Villages such as Egmond aan Zee, Petten and Callantsoog were forced to relocate either partially

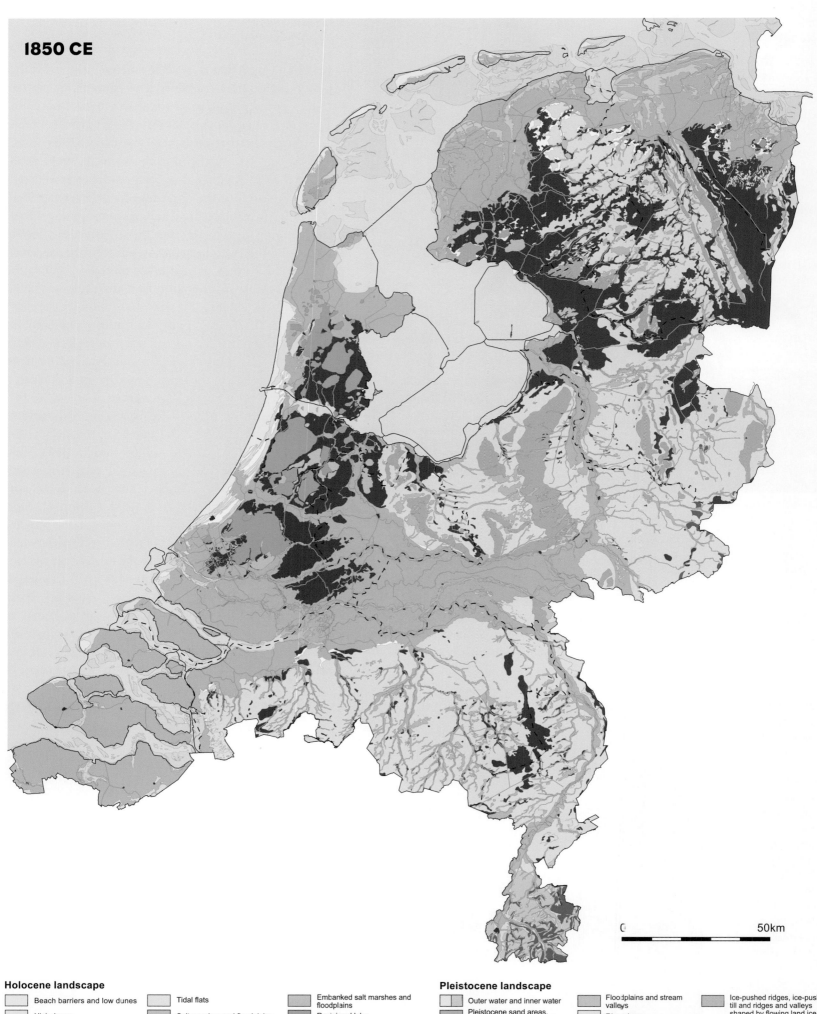

1850 CE

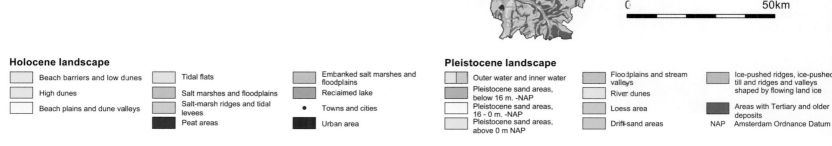

Holocene landscape

Beach barriers and low dunes	Tidal flats	Embanked salt marshes and floodplains
High dunes	Salt marshes and floodplains	Reclaimed lake
Beach plains and dune valleys	Salt-marsh ridges and tidal levees	• Towns and cities
	Peat areas	Urban area

Pleistocene landscape

Outer water and inner water	Floodplains and stream valleys	Ice-pushed ridges, ice-pushed till and ridges and valleys shaped by flowing land ice
Pleistocene sand areas, below 16 m. -NAP	River dunes	
Pleistocene sand areas, 16 - 0 m. -NAP	Loess area	Areas with Tertiary and older deposits
Pleistocene sand areas, above 0 m NAP	Drift-sand areas	NAP Amsterdam Ordnance Datum

50km

In Holland in particular, peat cutting was so intensive that there was soon almost no peat left above ground. People then turned to 'peat dredging' (*slagturven*), which involved removing peat layers to a depth well below the groundwater table. This left large lakes that were several metres deep. Their shores crumbled away still further during storms, posing a major threat to the surrounding area. The peat-dredgers were officially required to drain the lakes immediately after their work was done, but in practice this usually came to nothing. All the same, pumping the lakes dry was the only effective solution. At first, windmills were used for this purpose (the Schermer, Beemster and Purmer are just a few examples), but in the nineteenth century steam engines were also deployed. A high point was reached in 1849-1852 with the draining of the Haarlemmermeer. Three steam pumping stations were used, including the Cruquius, the sole survivor today. In 1850 there were only a few large peat lakes left in South Holland. But there was still no solution to the threat posed by the Zuiderzee, a large inland sea in the heart of the country.

In the river area too, people were able to control the adverse effects of natural processes. The courses of the major rivers were increasingly fixed. The Rhine, Waal, IJssel, Lek, Merwede and Meuse had long enjoyed relative freedom within their winter embankments. That freedom was curtailed for safety reasons, and later also for shipping. From the eighteenth century, islands were removed and groynes were built at right angles to the current, directing the water into a single channel. One major problem was that the embankments were often breached, especially when accumulated ice prevented the water from being discharged. These breaches caused characteristic 'scars' in the dyke, known as *wielen*. To combat this problem, large overflow channels (*overlaten*) were built between the rivers in the eighteenth and nineteenth centuries to divert the water in times of flood. In

or entirely. The Wadden islands weren't stable either; villages there had to be moved because of the threat of erosion.

Inland, peat reclamation continued unabated after 1500. In 1850 the peatlands had shrunk significantly and large areas of peat were still only found in the Ronde Venen (between Amsterdam and Utrecht), southeast Groningen and the Peel. There too, the extraction of this 'brown gold' was in full swing. Several decades later, the peat bogs had largely vanished from the Netherlands.

FIG. 70 The Vollenhoven estate in De Bilt is a fine example of the many country homes that were built in the eighteenth and nineteenth centuries. Because of the originality and integrity of its main building, outbuildings and English landscape garden, Vollenhoven is regarded as the jewel in the crown of the Stichtse Lustwarande area.

only a few cases did the overflow channels have the desired effect, such as the Beerse and Baardwijkse overlaat in the Meuse between Grave and Waalwijk. Other overflow channels, however, such as the Lijmerse overlaat, which was designed to divert water from the Rhine to the IJssel, proved less effective.

In 1850 people didn't yet have *all* geological processes under control. The inhabitants of the high Netherlands and the coastal dunes had to sit idly by while their villages and agricultural areas were threatened by drift sand. These drift sands were the result of human intervention. The farmers used the wasteland to graze sheep and to cut sods. The sheep manure and sods were then used to maintain the fertility of their arable land. But over-intensive use turned these wastelands into sandy plains. Humans were therefore responsible for two new types of soil in the high Netherlands: essen soils, which they cultivated, and – unintentionally – drift sands.

Despite these local problems, between 1500 and 1850 people had gained still greater control over the landscape, enabling more intensive habitation and agricultural exploitation. The pattern of existing towns and villages was consolidated and the period from about 1525 to 1675 was even marked by a fairly sharp rise in the urban population. In 1675 about half of the population of the western Netherlands lived in cities – an exceptionally high percentage by the European standards of the time.

This period also witnessed the emergence of the first industrial landscapes. The best-known example is the Zaanstreek, where hundreds of sawmills, oil mills, hulling and paper mills operated. The 'Golden Age' left the biggest mark on the landscape of the western Netherlands, but elsewhere too the growing prosperity was clearly in evidence. On the periphery of the Veluwe, streams were artificially extended in an upstream direction. The purpose of these *sprengen* was to improve water supply for the local paper industry. And wherever possible, artificial watercourses were dug in this region, once again for the benefit of the paper mills.

Population growth and an increase in trade led to a rise in the number of waterways. In the seventeenth century the cities organised the excavation of a network of tow-canals, a system that survived until trains took over the role of the horse-drawn boats in the latter half of the nineteenth century. The first half of that century also saw the excavation of several major canals for the transport of goods. Examples are the North Holland Canal from Den Helder to Amsterdam (1824) and the South Willem's Canal (1826), which linked the industrial area around Liège to the Meuse at 's-Hertogenbosch. The economic prosperity was reflected in the construction of large numbers of country estates, including at 's-Graveland, on the Vecht, along the inner edge of the dunes from Beverwijk to Haarlem, at Wassenaar and east of Utrecht (the Stichtse Lustwarande).

In the period 1500-1850 several landscape interventions were made in the context of the country's defence. In the Middle Ages forts and city walls had sufficed to keep out the enemy but at the end of the sixteenth century, during the war with Spain, the idea of a line of defence took hold. This involved a series of defence works with intermediate zones that could be flooded in times of peril. In 1815 a start was made on the New Dutch Waterline, the successor to the old waterline of the seventeenth century.

The degradation of the undeveloped landscape had consequences for the country's fauna. In the early eighteenth century the last beaver was killed, wolves were all but eradicated and the last crane was spotted. Red deer did make a reappearance, but they had been reintroduced for hunting in the sixteenth and seventeenth centuries. The agrarian landscape was increasingly enhanced by a rich variety of food crops – including the potato imported from South America – and different breeds of cattle, sheep, goat, horses and poultry. Turkeys, for example, were introduced from South America. Close relations with the colonies were also reflected in the flora. Exotic plants were imported from all around the world and planted at country homes and in botanical gardens.

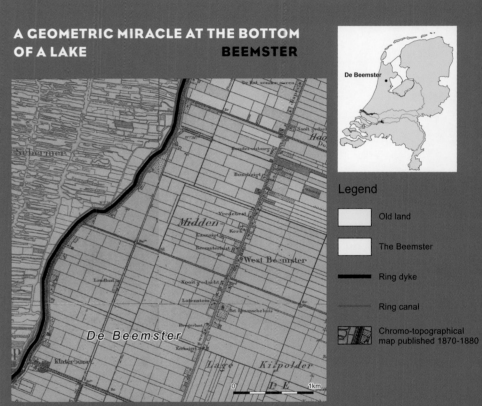

FIG. 71 Location map of the impoldered Beemster.

The Beemster was one of many lakes in Holland that were created when a small peat river, in this case the Bamestra, was enlarged. The lake was drained in 1612, a private initiative by mainly Amsterdam merchants and city administrators wanting to invest their rapidly expanding fortunes. The bottom of the Beemster, some 3.5 metres below sea level, was composed almost entirely of heavy clay. It was the investors' hope that this land would have excellent agricultural potential.

What was so unusual about the Beemster was its strict, orderly geometric layout, the product of both a rational engineering and land surveying approach and the conscious application of aesthetic principles. The tight layout would eventually provide space for some 350 farmhouses (*stolpboerderijen*) and more than fifty country estates. The latter disappeared again in the eighteenth and nineteenth centuries, however, because of declining prosperity.

The Beemster was initially drained and kept dry with the aid of 43 windmills. The water regulation had to be adjusted several times over the years to ensure that the lowest parts of the polder stayed permanently dry.

The Beemster polder has been awarded a place in the Canon of Dutch History for its landscape and architectural values. It also features on the UNESCO world heritage list.

2000 CE
A COUNTRY CREATED BY PEOPLE

FIG. 72 The large modern city of Eindhoven. The landscape is entirely built up. While the roads in the old city centre in the middle of the photo still follow the contours of the old landscape, the more recent parts of town beyond the centre have been laid out in a rational pattern.

Humans were already the dominant geological factor in the Netherlands in 1850, but by the year 2000 there was almost no part of the country that hadn't been shaped by human hand. The Zuiderzee Works, the Delta Works and the Maasvlakte (Port of Rotterdam) stand out immediately on the map, as does the spectacular rise in urbanisation. Roads and railroads take up an increasing share of the surface area. Less conspicuous but certainly just as profound is the impact of the large-scale land consolidation and reclamations of the twentieth century. Human intervention could even be seen below ground.

In about 1850, the natural processes that for millennia had both built up and broken down the soil of the Netherlands were largely brought under control – but not entirely. In 1916 the Zuiderzee dykes in North Holland collapsed under the force of a powerful winter storm. The resulting flood then prompted the launch of the Zuiderzee Works, whereby the Zuiderzee was closed off from the Wadden Sea with the construction of the Afsluitdijk in 1932 and parts of the IJsselmeer were reclaimed. History repeated itself in 1953 when an even larger flood struck the Zeeland and South Holland delta and more than 1800 people met their death. In the wake of this disaster, the Delta Plan saw to the raising of dykes, the construction of storm surge barriers and the shortening of coastlines. These human interventions had an unprecedented impact. The Zuiderzee turned into fresh water and became the IJsselmeer, with a completely different flora and fauna. These days the lake plays a significant role in fresh-water management.

Large-scale changes were also made in Zeeland and South Holland. The Haringvliet, Grevelingen and Volkerak were

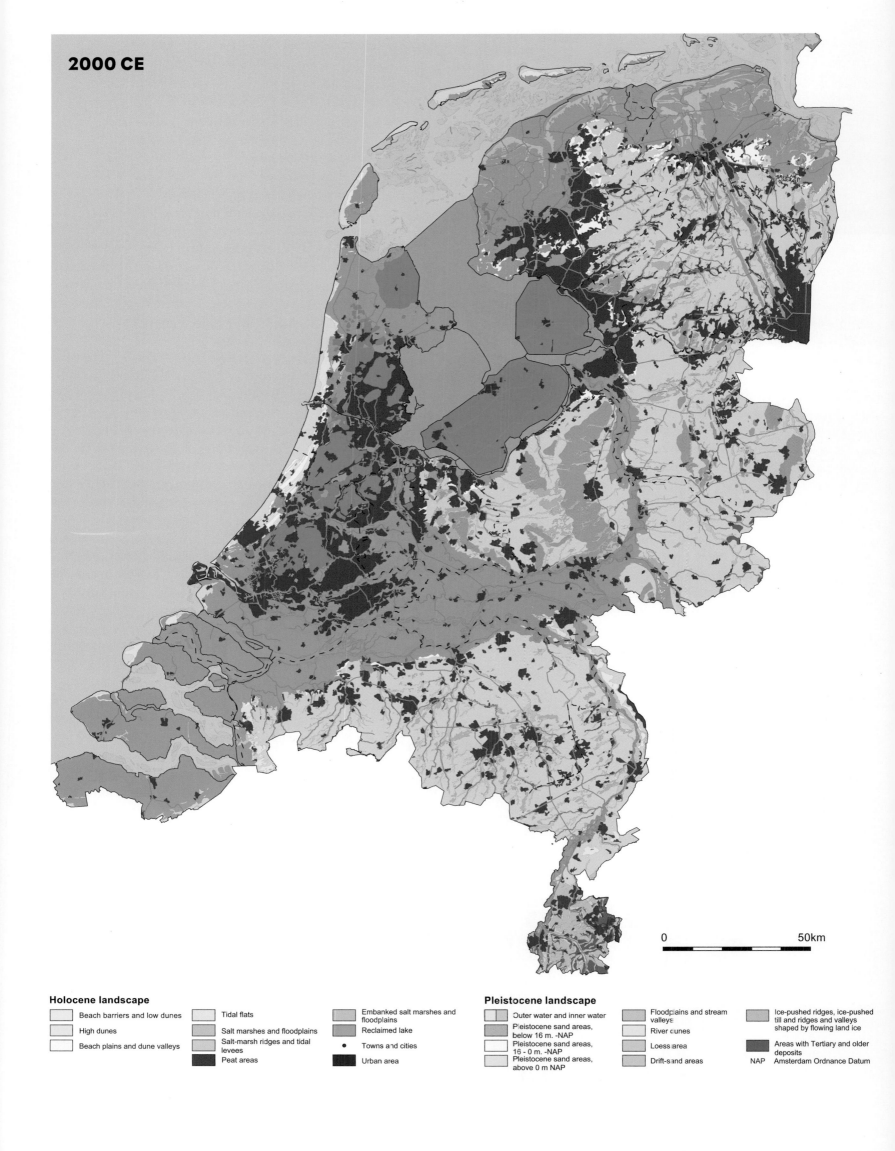

2000 CE

Holocene landscape

- Beach barriers and low dunes
- High dunes
- Beach plains and dune valleys
- Tidal flats
- Salt marshes and floodplains
- Salt-marsh ridges and tidal levees
- Peat areas
- Embanked salt marshes and floodplains
- Reclaimed lake
- Towns and cities
- Urban area

Pleistocene landscape

- Outer water and inner water
- Pleistocene sand areas, below 16 m. -NAP
- Pleistocene sand areas, 16 - 0 m. -NAP
- Pleistocene sand areas, above 0 m NAP
- Floodplains and stream valleys
- River dunes
- Loess area
- Drift-sand areas
- Ice-pushed ridges, ice-pushed till and ridges and valleys shaped by flowing land ice
- Areas with Tertiary and older deposits
- NAP   Amsterdam Ordnance Datum

0   50km

FIG. 73 The *Waterway*, a trailing suction hopper dredger, in action. Jumbo vessels like this are currently being used to create Maasvlakte 2. The *Waterway* can dredge up sand at depths of up to 28 metres.

dammed-up completely. One consequence was that the Biesbosch now experienced almost no tidal movement and the characteristic freshwater tidal zone gave way to a freshwater delta with swamp forests of poplar and willow.

A decision was made in 1976 to build the Eastern Scheldt storm surge barrier in order to ensure the tides could still come into the Oosterschelde. This would enable the preservation of the Oosterschelde's natural mudflat and salt-marsh environment, with its tidal ebb and flow. This feat of Dutch engineering has proven only partially effective: the remaining tidal movement is not quite enough to preserve the former environment.

Another striking development on the coast was the creation of the Maasvlakte, where an industrial zone has been built on new land stretching far into the North Sea.

The construction of all these major works has affected the coastal current. Following the partial damming of the Zeeland coastal inlets, much less seawater flows in and out with the tide. The coastal sand shoals along the Zeeland coast adopted to the new situation. New sandbars were formed parallel to the coastline, known as the 'fore-delta'. The construction of the Maasvlakte has interrupted the current that ran parallel to the coast. Parts of the coastline of Goeree-Overflakkee

FIG. 74 After World War Two, millions of single-family dwellings were built in the Netherlands, like the one here in Vathorst, Amersfoort. The demand for housing continues to be high.

and Voorne are now accreting. The building of the Afsluitdijk affected the current in the western Wadden Sea: the tidal channels shifted, as did the sandflats that are dry at low tide. The same happened on a smaller scale in the eastern Wadden Sea following the impoldering of the Lauwerszee.

Aside from these human interventions, erosion has been outstripping sedimentation along the coast since Roman times. As a result, the coast has receded several kilometres and is severely weakened in a number of places. People have also combated this threat by replenishing the shore face and beaches with sand to reinforce them ('sand nourishment'). These days the Ministry of Infrastructure and Water Management (Rijkswaterstaat) takes advantage of the natural current and surf action. Sand is extracted from further out at sea and dumped on the shallow seabed just off the coast; the current and surf then bring it to the beach.

By 2000 the major rivers were fixed by groynes (insofar as this hadn't been done already by 1850) and parts of some rivers were even equipped with stone banks. Thanks to major works carried out on the rivers and the further raising of the dykes, there was a much smaller risk of dyke breaches. At Gorinchem, where the Meuse discharged into the Waal, and further upstream at Heerewaarden, where the two rivers also came together, danger had lurked on many occasions at high water. Between 1850 and 1950 these rivers were separated. The excavation of the Nieuwe Merwede and the Bergsche Maas had accelerated the discharge of flood waters to the sea. Because of the reduced risk of river embankments being breached, the rather unsuccessful overflow channels could be closed. But this didn't mean that all risk of flooding had been eliminated. Water levels in the floodplains could still rise rapidly, as was shown by the major floods along the Meuse in Limburg (1993 and 1995) and the narrowly-averted Rhine flood disaster (1995). These events, together with predictions about a more capricious climate, have made policy-makers appreciate the importance of more natural floodplains and other flood catchment areas. Large-scale interventions are being made as part of the Maaswerken project and the Room for the River programme.

The extensive peatlands that were once so characteristic of the Netherlands have almost completely vanished. What remains is for the most part managed as nature reserves. This does, however, entail keeping the groundwater table artificially high in order to prevent the peat from disappearing through oxidation.

The problem of drifting sand in the high Netherlands was solved in the twentieth century. The advent of artificial fertiliser offered a solution: farmers no longer had to rely on sheep manure and sod manuring. The driftsand areas could be reforested, which happened on such a major scale, especially after 1920, that there are almost no driftsands (or heathland) left.

One of the most far-reaching changes in the landscape has been the adjustment of the groundwater table for agricultural purposes. Almost no area in the Netherlands still has a natural groundwater regime. Virtually everywhere, fertilisation has made the shallow groundwater much more nutrient-rich. This is even the case in many nature reserves, through the precipitation of nutrients such as ammonia. Once-common plants that thrived in nutrient-poor conditions can only survive because of concerted efforts to remove nutrients from the soil.

Between 1850 and 2000 the growing impact of technology on the landscape went hand in hand with an enormous population upsurge (from almost five million in about 1900 to some 17,250,000 at the end of 2018). This period saw an explosive growth in the built environment, a process that was further reinforced by declining household size and a sharp rise in the amount of living space per household.

In addition to homes, the built environment was taken up by major industrial complexes and infrastructure works. The network of waterways increased in density after 1850. Examples of late nineteenth-century canals are the North Sea Canal, the Merwede Canal (later incorporated into the Amsterdam-Rhine Canal), the Nieuwe Waterweg, the Overijsselse canal system and the Eems Canal. The railway network also grew rapidly after 1850. And with the rise of the automobile, there was also an enormous increase in the number of roads in the twentieth century. In hundreds of places, Dutch waterways were bridged or tunnelled under to facilitate through traffic. Schiphol, one of the biggest and busiest airports in Europe, was built in the centre of the largest reclaimed lake in Holland.

The construction of permanent lines of defence, intended from the seventeenth century onward to provide greater military security, continued. The New Dutch Waterline was extended and the Defence Line of Amsterdam (Stelling van Amsterdam) was added. Further additions were the Stelling van het Hollands Diep en Volkerak and the Stelling van der monden der Maas en van het Haringvliet. During World War Two the German occupying forces built a whole series of concrete fortifications – the *Neue Westwall* or *Atlantikwall* – along the Dutch coast. The postwar IJssel line, designed to ward off a Soviet invasion, would be the last line of defence: long-range rockets and planes had rendered deliberate flooding futile.

The appearance of the countryside changed profoundly, especially after World War Two. Farming has increased enormously in scale. Land consolidation projects have sometimes changed much of the Dutch landscape beyond recognition. Modern agriculture has made full use of pesticides and fertilisers; in combination with the artificial lowering of the groundwater table, they have dramatically reduced the biodiversity of wild meadow flora.

Livestock have become much more standardised as a result of rationalised farming: the many breeds of the past have vanished from the landscape and have often been replaced by exotic breeds. At the end of the twentieth century the country's millions of pigs and chickens were practically invisible because they never left their sheds and sties; only the cows were still outside. Paradoxically, the upshot of all this is that these days the sandy areas, with their enormous pigsties, are dominated by horses for recreational use. Here too the diversity of agricultural crops has fallen sharply, with silage maize now the dominant crop.

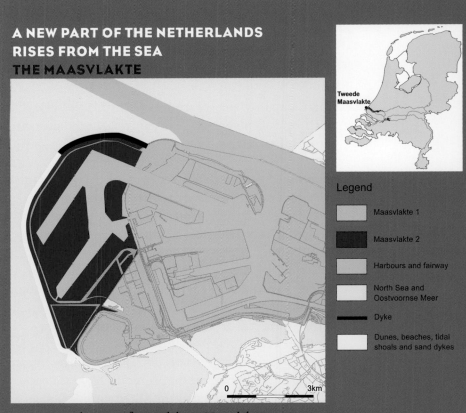

A NEW PART OF THE NETHERLANDS RISES FROM THE SEA
THE MAASVLAKTE

Legend
- Maasvlakte 1
- Maasvlakte 2
- Harbours and fairway
- North Sea and Oostvoornse Meer
- Dyke
- Dunes, beaches, tidal shoals and sand dykes

FIG. 75 Location map of Maasvlakte 2. Maasvlakte 2 was completed in 2015.

In the 1960s the city of Rotterdam made a start on construction of the Maasvlakte. The Port of Rotterdam had become too small for ships and their ever-increasing draughts. But instead of deepening the Nieuwe Waterweg, Rotterdam built a completely new port area *in* the North Sea.

The Maasvlakte, completed in 1973, is in fact one giant terp. It consists of sand that has been dredged from the North Sea floor by large hopper dredgers and brought to the Maasvlakte in a process called 'rainbowing' (in which a dredging ship propels sand in a high arc to its new location). Because the ports of the Maasvlakte are in open connection with the North Sea, the 'terp' needed to be very high: it had to remain unsubmerged at times of extremely high water. The quantities of 'rainbowed' sand were therefore enormous. 'Maasvlakte 2' alone, the extension of the first Maasvlakte, required 240 million cubic metres of sand.

Because the Maasvlakte extends a little into the North Sea, it has altered the sea current and caused large quantities of sand to build up off the coast of Voorne-Putten. An interesting detail is that the dredged-up sand contains many fossils, including those of mammoths. Mesolithic spearheads, once at the bottom of the North Sea, have been found in the sand of Maasvlakte 1.

In the twentieth century, this sparked a counter-movement and a growing need for a nature that was 'unspoiled'. In 1906 Vereniging Natuurmonumenten, the Dutch nature preservation society, purchased the Naardermeer and established the first 'nature reserve' there. Many more would follow. Few people realise that we are preserving a 'nature' that we ourselves have created. Despite these moves, animal species such as wolves, otters and storks disappeared from the landscape. The number of fish species declined dramatically as a result of river and stream pollution. And yet the tide has turned somewhat since the dawning of the new millennium. Thanks to environmental measures, fish stocks have risen again, beavers and storks have been released and cranes, egrets, wild cats and wolves have been sighted once more. And coypus, introduced from South America for their fur, are now part and parcel of the Dutch landscape following escapes from fur farms.

GLOSSARY

Only the main references are included in this glossary.
Dutch equivalents are shown in italics. Figure numbers are
shown in brackets.

abandoned river bed ridge *stroomrug* **p. 21 (13B), 51** Fairly
elevated ridge in a river delta that is the remnant of a
former, abandoned river course. The levees of the abandoned river bed ridge are higher than the surrounding
river basin.

accretion *verlanding* **p. 19 (12)** The closing over of a lake,
pond or former river course as a result of peat formation
or silting-up.

Ahrensburg culture *Ahrensburgcultuur* **pp. 38-9** The
culture of late Palaeolithic hunter-gatherers who lived
from c. 10,700 to c. 9700 BCE, during the Younger Dryas,
the last cold spell before the beginning of the Holocene.
The culture is named after the Ahrensburg-Stellmoor find
site, 25 km northeast of Hamburg in the German state of
Schleswig-Holstein. Small tanged points are characteristic of the Ahrensburg culture.

anastomosing river *anastomoserende rivier* **pp. 20-21** Type
of lowland river whose individual channels branch and
rejoin but, unlike meandering rivers, do not change
position or break out. The Rhine and Meuse tended to
anastomose downstream, where there was a low flow rate,
high sedimentation and where raised bogs occurred.
These factors combined made it unlikely that these rivers
would erode or burst their banks.

basin storage capacity *komberging* **p. 48, 68** The capacity of
a **tidal basin** behind **beach barriers** to store incoming
sea water (see **tidal volume**) at high tide. Because a tidal
basin has fairly narrow openings, the effect of tidal ebb
and flow within the basin is weaker than out at sea,
especially if the openings are very narrow and the tidal
basin is very large. The tidal basin is not completely full
when the sea water reaches its highest point. Conversely,
the tidal basin is not fully drained when the sea water is at
its lowest point. Researchers wishing to establish the
height of relative sea level need to take this phenomenon
into account.

beach barrier *strandwal* **p. 14, 19 (12), 22, 25 (16), 32 (22)**
An elongated aeolian sand ridge that runs more or less
parallel to the coastline. Beach barriers are formed when
sand accumulates on the beach and is transported by
wind to the barrier ridge.

Bell Beaker culture *klokbekercultuur* **p. 51** A culture of
sedentary farmers (c. 2400-1900 BCE) who derive their
name from pottery beakers resembling an inverted bell.
The Bell Beaker culture was scattered across much of
Europe.

bog *hoogveen* **p. 68** Peat that has formed above the original
groundwater table from plants that depend on rainwater
for their growth (usually *Sphagnum* moss). This term is
synonymous with oligotrophic peat.

braided river *vlechtende rivier* **pp. 20-21 (13A)** A river with
many channels in a broad, sandy river course in which the
channels constantly shift.

burial mound *grafheuvel* **p. 48, 51, 55, 59** An artificial
mound, often made of sods, clay or sand, in or beneath
which one or more graves are located. The graves may
contain either inhumation or cremation burials. Burial
mounds are also known as barrows or tumuli (sing.
tumulus).

byre house *woonstalboerderij* **p. 55, 59** A farmhouse building typical of northwestern Europe, in which people and
livestock lived under one roof. This style of farmhouse
developed in the Middle Bronze Age and predominated in
farming communities until well into the twentieth
century.

Carolingians *Karolingers* **p. 71** The descendants of Charles
Martel (676-741). Originating from the area between
Namur and Liège, they formed a Frankish dynasty that
rose to power in the seventh and eighth centuries. They
were the de facto rulers in the realm of the Merovingian
kings. Pepin the Short was crowned king of the Franks in
751. The most famous Carolingian was his son, Charlemagne (747-814).

carr *broekbos* **p. 23 (15A)** A low-lying waterlogged and
wooded terrain, often in areas that stay wet through
seepage. Carrs are also found along natural streams and
rivers that flood, and are submerged for a long period in
winter. Characteristic trees include alder and willow.

carr peat *broekveen* **p. 22 (clayey, non-clayey)** Peat that is
composed of the remains of carr, primarily alder. Carr
peat is often clayey, especially if it is formed in the
floodplains of rivers and streams. The term is often used
as a synonym for wood peat but this is factually incorrect.
Wood peat refers to all kinds of peat that contain tree
remains.

'Celtic' fields *raatakker* **p. 59** A patchwork of small, rectangular plots, separated by low banks. Waste from the
reclamation, weeding and harvesting of the fields was
dumped on the periphery of the field plots, in this way
creating the banks. Despite the regular application of
manure, the fields eventually became depleted. This may
explain why the banks were pressed into agricultural use
in their place. The word Celtic appears in inverted com-
mas because 'Celtic' fields in fact have nothing do with the
Celts.

clay (marine, river, humus) *klei (marine, rivier, humeus)*
passim A cohesive, fine-grained soil type that develops
plastic properties when water is added. In the Netherlands, clay is formally defined in terms of the proportion
of particles smaller than 2 μm (0,002 mm). This gives a

further classification into heavy or light clay, especially in agriculture, where it refers to the workability of clay soils. The heavier the clay, the greater the proportion of particles smaller than 2 μm. The name given to the type of clay, such as marine clay and river clay, may also indicate how the clay was deposited. Marine and river clay often contain a certain percentage of organic material (plant and animal remains). If that percentage is high, it is called humus clay.

coastal dune *kustduin* **p. 10, 19, 44** A **dune** along the coast. Coastal dunes are formed from sand blown up from the beach, where it was carried by the current and surf.

coastal peat bog *kustveenmoeras* **pp. 25-26, 40** A bog behind closed beach barriers, where peat forms because of the high groundwater level. This coastal peat can eventually develop into a raised bog several metres in height.

coversand *dekzand* **p. 33** Sand deposited by the wind at the end of the Pleistocene, during the last ice age when the Netherlands had a very dry polar climate. Sand was easily transported by the wind in the polar desert. Coversand still occurs on the surface in large parts of the northern, eastern and southern Netherlands. In the western Netherlands the coversand was overlain by other layers (peat and clay) or washed away during the Holocene. The sand has a fairly constant particle size, averaging 150-210 μm (0.150-0.210 mm).

coversand hill *dekzandkop* **p. 33** Windblown sand that forms a low dune rising above the surrounding coversand plain. In coversand areas that were overlain with peat and clay during the Holocene, these low dunes often long survived as hills rising above the peat, making them ideal settlement sites in prehistory.

creek *kreek* **p. (39), 67** A small tidal channel in a tidal basin, often situated between salt marshes.

crevasse **(13) p. 55** Although most commonly used to refer to a deep crack in a glacier, crevasse is also the term for a crack or fissure in a riverbank as the result of erosion. When water levels are high, water from the river flows through the crevasse into the floodplains. The term crevasse can refer not only to the fissures itself but also to the characteristic deposits behind them (in effect a mini-delta extending towards the floodplain). A crevasse can also mark the preliminary stage of a new river course.

cultivation of new land *ontginnen* **pp. 71-72, 79** Making a comparatively natural landscape suitable for human use by altering the natural conditions; also described as putting wasteland into cultivation. Examples of reclamation in the Netherlands are the impoldering and parcelling of salt marshes, the cultivation of peatlands, the draining of lakes and ponds, deforestation and the cultivation of large tracts of heathland on the higher sandy soils (made possible with the advent of artificial fertilisers).

dam p. 24, 72 During the Middle Ages, dams were constructed on a large scale on the side channels of rivers, creeks or tidal inlets in order to stop the inflow of water. Dams were constructed on a large scale during the Middle Ages.

donk p. 42, 47 The top of a **river dune** that rises above the surrounding Holocene deposits. Donks are mainly found in the Rhine and Meuse deltas and were ideal places to settle in early prehistory.

drift sand *stuifzand* **p. 33, 79, 83** General term for sand deposited by the wind, often used in the Netherlands to describe sand that began drifting in the overgrazed heathlands from the Middle Ages onwards.

dune *duin* **p. 33, 50 (29)** A hill of windblown sand. The dunes on the Dutch sea coast, which can rise to a height of more than 50 m, were formed from about 1000 CE. There are also dunes inland, in drift-sand areas that developed from the Middle Ages as a result of the overgrazing of heathland. The **river dunes** along the major rivers were formed at the end of the Pleistocene when sand was blown from the dry parts of braided rivers. Within geology, the term 'dune' also includes sand ripples with a crest height of more than 5 cm.

dyke *dijk* **p. 19, 74, 76** Dykes and embankments were built to stop river or sea water from flooding the hinterland. In the Netherlands, dykes were built on a large scale from the eleventh century onwards, both along the coast and in the river region.

es (pl. essen) or enk p. 75 A large, open cultivated area, part of which was originally in communal use. Essen, which are found on the higher sandy soils in the northern, eastern and southern Netherlands, were formed from the Middle Ages. They are always higher than the surrounding land and were mainly used as arable land. There is often a clearly marked transition to the surrounding landscape, sometimes in the form of border planting. Essen were not built on, and thoroughfares ran around their periphery rather than through or across them. Essen often consisted of several blocks of reclaimed land, which until recent times were subdivided into numerous narrow strips of land.

essen (or plaggen) soil *esdek* **p. 83** The layer, at least 50 cm thick, of arable soil covering an es, formed as the result of the prolonged application of a mixture of manure and sandy sods (sod manuring).

estuary *estuarium* **p. 19** The mouth of a river where it broadens into the sea. The tide penetrates far into the estuary, causing saline and fresh water to intermix. Estuaries often contain **salt marshes** and **tidal flats**.

eutrophic *eutroof* **p. 22, 25** Rich in nutrients. This term is often used to describe a type of **peat**. Eutrophic peat is peat that has formed in nutrient-rich water. Wood peat and reed peat are examples of eutrophic peat types.

floodplain *komgebied* **p. 19, 20 (13b)** Low-lying area adjacent to a river that is inundated in times of flood. The soil in the floodplain often consists of clay that is deposited during the flooding. In the Netherlands, the building of dykes means that the floodplains are no longer flooded. The area between the summer and winter dykes (*uiterwaarden*) provide only a fraction of the room once available to the river at high water.

Frankicisation *frankisering* **p. 71** The sociopolitical, cultural and religious process (at times peaceful, at times violent) whereby tribal groups were absorbed into the early polities of the Merovingian and Carolingian kings (fifth to the ninth century CE).

freshwater tidal zone *zoetwatergetijdengebied* **p. 78** Part of the river mouth in a delta or estuary that is influenced by the tide but which still contains fresh water. In times of extremely low river discharge or **spring tide**, sea water can temporarily penetrate into parts of the freshwater tidal zone. Before the Haringvliet was cut off from the sea, the Biesbosch was an example of a freshwater tidal zone.

Funnelbeaker culture *trechterbekercultuur* **p. 50, (38), 51** The collective name for a large number of Middle Neolithic farming communities in southern Sweden, the northern Netherlands, northern Germany and Poland. The Funnelbeaker culture in the Netherlands (c. 3400-2900 BCE) belonged to what is known as the 'western group'. Most of the findspots in the Netherlands occur in the Pleistocene sandy soils in the eastern Netherlands, and on the Hondsrug sand ridge in particular. Research into the Funnelbeaker culture has traditionally emphasised the pottery from megalithic tombs (Dutch *hunebedden*). Other characteristics, such as settlement locations, have been less well researched.

groyne *krib* **p. 82** A stone construction in the form of a short dam built at right angles to a river bank or beach in order to prevent erosion by the current.

Holocene *Holoceen* **pp. 10-13** The most recent geological epoch, from 9700 BCE to the present. The Holocene is characterised by a fairly equable, warmer climate than in the preceding ice age.

ice age *ijstijd* **p. 11, (3), 13, 15 (5)** Any period in Earth's history in which the continents were partly covered by large ice sheets and temperatures were also very low in areas that have a temperate climate today. Ice ages occurred in different periods of Earth's history. The most recent series of ice ages, which also led to major cooling in the northern hemisphere, began about 2.6 million years ago. The Holocene, the epoch in which we currently live, is a warm period between two ice ages.

ice-pushed ridge *stuwwal* **p. 13 (4), 33** An elevation in the landscape created by an ice sheet. An ice-pushed ridge is composed of sediment that has been pushed forwards and sideways by the ice sheet. The ice-pushed ridges in the Netherlands date from the penultimate ice age (Saalian) and are approx. 150,000 years old.

intertidal zone *intergetijdengebied* **pp. 60-63** The area in tidal basins and estuaries between mean high and low water levels. This environment is dry twice a day (at low tide) and is flooded twice a day (at high tide).

isostasy **pp. 15-17** *isostasie* A geoscience term referring to a state of gravitational equilibrium of the solid upper layer of the earth (the lithosphere, consisting of the crust and the rigid upper mantle) that 'floats' on the viscous part of the mantle below it (the asthenosphere). By analogy with a ship that sinks lower into the water as it is loaded with cargo, isostasy describes the principle of buoyancy, whereby a load on the lithosphere causes it to sink as the rock in the asthenosphere 'flows away' under the load. If the load is removed, the reverse process occurs and the lithosphere floats up again. Istostasy can be observed in a tectonic context, in which a mountain range may exert pressure, but it is also critical for explaining relative sea-level movements during ice ages (glacio-isostasy). Land ice that forms during ice ages pushes down the lithosphere, causing the Earth's crust at the edges of the ice cap to rise up as the viscous rock in the asthenosphere flows to areas without an ice load. When the ice cap melts, the reverse process occurs: the area once covered in ice rises and the adjacent areas sink. Because of the sluggishness of the rock flow in the asthenosphere and the rigidity of the lithosphere, isostatic equilibrium does not occur at the same pace as the formation and melting of ice caps: the vertical movement of the Earth's surface still lags behind the disappearance some 10,000 years ago of the northern hemisphere's ice caps from the last ice age. The formation and melting of Arctic sea ice do not have a glacio-isostatic effect because it does not lie atop the lithosphere, but floats in the Arctic Ocean.

jet (or 'black amber') *git* **p. 51** An organic black or dark-brown amorphous mineral created under pressure from fossilised wood. Jet can be polished to a bright lustre.

lagoon *lagune* **p. 19** A shallow body of brakish or salt water behind beach barriers which has a small or indirect connection to the sea. The term is sometimes used as a synonym for tidal basin, but this is incorrect. Whereas large parts of a tidal basin are dry at low tide, a lagoon is not influenced by tides and is always filled with water.

land reclamation *landaanwinning* **p. 80** The reclamation of land from the sea by allowing it to silt up and then impoldering it, or by damming inlets and then gradually draining them. The first method was used to reclaim large parts of Zeeland, Friesland and Groningen from the sea, while the Wieringermeer, Noordoostpolder and Flevo-polders are examples of drained parts of the former Zuiderzee seabed.

levee *oeverwal* **p. 55** A low bank alongside the channel of a river. It is formed naturally by the river, which deposits sediment there in times of flood.

limes **p. 66, 71** The modern technical term for the fortified border defences of the Roman empire. In the Roman province of *Germania inferior* (in the Netherlands and neighbouring Germany), the *limes* consisted of an artificial road and forts, watchtowers, river ports and landing stages along the Rhine (Oude Rijn, Kromme Rijn and Nederrijn). The *limes* was created in the middle of the first century CE and survived, with interruptions of varying length, until the early fifth century. As well as being a system of defence works along the imperial border, the *limes* was also an important protected communication route.

Linearbandkeramik (LBK) culture *bandkeramische cultuur* **pp. 46-7** A culture of farmers from central Europe, named after their distinctive pottery style (*Linearbandkeramik* in German). The culture probably arose in southeastern Europe. Between c. 5400 and 4900 BCE LBK colonists spread rapidly from the Great Hungarian Plain to the loess regions of Slovakia, Czechia, Austria, Germany, Poland, eastern France and the Low Countries. The LBK culture was characterised by the introduction of agriculture, livestock farming, large monumental houses, adzes and pottery decorated with incised bands of geometric decorations. In the Netherlands, remains of LBK settlements dating from between c. 5300 and 4900 BCE are found only in southern Limburg.

loess *löss* **p. 72, 73, 77, 81, 85** Loam deposited by the wind (with a particle size ranging from 2 to 50 μm, or 0.002 to 0.05 mm). In the Netherlands, loess occurs in South Limburg, along the Veluwezoom near Arnhem and behind the Nijmegen ice-pushed ridge at Groesbeek. It was deposited there during the coldest part of the last ice age, when the Netherlands had a polar desert climate.

meandering river *meanderende rivier* **p. 20** A river that meanders through the landscape, with constantly shifting bends.

Merovingians *Merovingers* **p. 71** The descendants of the mythical king Merovech of the Salian Franks from the time of the 'Migration Period'. The history of the Merovingian dynasty is well documented, from the accession of Childeric I (c. 457-481) to the downfall of Childeric III, who was deposed in 751.

Mesolithic (early, middle and late) *mesolithicum* **pp. 38-39, 42-43** Middle Stone Age. The Mesolithic began in c. 9000 BCE and ended in 5300 BCE in areas where Bandkeramik farmers settled. Elsewhere the hunter-gatherer lifestyle continued to predominate until 4900 BCE, or even later. The Mesolithic is therefore the period between the end of the last ice age and the arrival of the first farmers. Mesolithic peoples inhabited a densely forested landscape and supported themselves through hunting and gathering. The advent of agriculture only gradually brought an end to this way of life.

mesotrophic *mesotroof* **p. 22, 25** Moderately rich in nutrients. This term is often used to describe kinds of **peat**. Mesotrophic peat has developed in water that is moderately rich in nutrients. Sedge peat is a mesotrophic peat type.

Neolithic *neolithicum* **p. (19), (27)** Late Stone Age (c. 5300/4900-2000 BCE). This period in the Netherlands saw the transition from the Mesolithic hunter-gatherer way of life to a more sedentary lifestyle in which arable and livestock farming predominated.

oligotrophic *oligotroof* **pp. 22-23** Nutrient-poor. This term is often used to describe kinds of **peat**. Oligotrophic peat has developed in nutrient-poor water, especially rainwater. *Sphagnum* peat is an oligotrophic peat type.

overlaat or overflow channel **p. 82** In its narrow sense, an *overlaat* is a spillway or overflow channel, i.e. the lower part of a river's winter dyke, allowing the river to flood in a controlled way at high water. In its broad sense, *overlaat* refers to the inundation areas behind the overflow channel as well as the waterworks that control the floodwaters. A well-known *overlaat* in the Netherlands, which was not closed until the twentieth century, was the Beerse Overlaat along the Meuse River in North Brabant.

Palaeolithic (early, middle and late) *paleolithicum* **p. 27, 39** The Old Stone Age (in the Netherlands from c. 300,000 until about 9700 BCE), the earliest period in human prehistory. This period began in the Netherlands with sporadic habitation by Neanderthalers during relatively warm periods. From the beginning of the late Palaeolithic, c. 13,000 BCE, we can speak in terms of habitation by modern humans. The Palaeolithic ended with the fairly abrupt end of the last ice age and the beginning of the present warm age, the Holocene.

peat *veen* **p. 15-17, 21-26, (13-17), 34, 44** A generally brown to black soil type consisting largely of plant remains that show no or only slight decomposition under oxygen-free conditions. Peats are commonly classified according to the type of plant remains that they are made of. Peat may contain considerable quantities of sand or clay. In order to qualify as peat, a sandy soil in the Netherlands must contain at least 15% organic matter, while 30% is the minimum for a clayey soil.

peat oxidation *veenoxidatie* **p. 17** The rotting of peat as a result of exposure to oxygen. The drainage of peat areas has led to large-scale peat oxidation, which together with soil compaction results in subsidence.

peat pasture *veenweidegebied* **p. 74** An area, often with elongated pastures, on a peat substrate. Peat pasture is most commonly found in the Dutch river area, parts of the Groene Hart, North Holland and Friesland.

Pleistocene *Pleistoceen* **p. 12 (2), 27, 30 (21), 33, 36** The geological epoch before the Holocene. The beginning of the Pleistocene, which is established at 2.6 million years ago, coincided with the appearance of the first large ice caps in the Northern hemisphere. The Pleistocene was characterised by alternating ice ages and warmer periods. The Pleistocene ended in 9700 BCE with the onset of the Holocene, the warm period in which we now live.

polder **p. 17, 28, 76, 83** Areas of land that are periodically or permanently lower than the surrounding water, from which they are protected by embankments. Water levels within polders are artificially regulated. Initially, this was done with the aid of locks, handmills or horse-driven mills, later with windmills and later still with steam, diesel and electric pumping stations. Most of the low-lying parts of the Netherlands consist of polders.

radiocarbon (or ^{14}C) dating *^{14}C-methode* **p. 16** A method of determining the age of organic material using the radioactive ^{14}C isotope. Three different isotopes of the element carbon occur in nature: the stable isotopes ^{12}C (approx. 99 percent) and ^{13}C (approx. 1 percent) and the radioactive isotope ^{14}C, with a half-life of 5,736 years ($< 10^{-10}$ percent). This last isotope is constantly being created in the atmosphere by the interaction of neutrons from cosmic radiation with atmospheric nitrogen: ^{14}N + 1 neutron $\rightarrow$ ^{14}C + 1 proton. The relative proportions of ^{14}C and ^{12}C in the tissue of living organisms, acquired through breathing and eating, is equivalent to the relative proportions of these isotopes in the atmosphere. When an organism dies, these proportions begin to change in a predictable way through the decay of ^{14}C. The clock then starts ticking. After 5,736 years, half of all the ^{14}C will have disappeared, three-quarters will have gone after twice 5,736 years, and 99.9% after 57,360 years. The age of an organism, determined by the proportions of ^{12}C and ^{14}C, is expressed in years BP (before present), with the 'present' being 1950. This year was not chosen at random: it was in 1950 that the first hydrogen bombs were tested in the atmosphere and atmospheric ^{14}C levels rose sharply above their natural values. Radiocarbon datings are calibrated by comparing them with absolute datings derived from studies of growth rings in trees, whose age can be very accurately determined back to thousands of years ago.

raised bog *veenkoepel* **p. 24 (15E), 25 (16), 26** A raised bog is a dome-shaped peat bog that is higher than the surrounding area. Raised bogs develop when *Sphagnum* moss starts growing in the peatland. *Sphagnum* moss is not dependent on groundwater but can thrive solely on nutrient-poor rainwater. Because peat moss continues to retain rainwater after it dies, it does not decay. Over the centuries this process of peat formation created peat domes, or raised bogs. Since the Middle Ages all the raised bogs in the Netherlands have been excavated for use as fuel. Finland and the Baltic countries still have large raised peat domes like the ones that once existed in the Netherlands.

reclaimed lake *droogmakerij* **p. 34, 87** A former lake or pond that has been pumped dry. The draining of lakes (the result of peat dredging) became possible with the advent of the hollow post mill (*wipmolen*) in the fifteenth century. The invention of the steam engine meant that larger lakes could also be drained. The polders in the reclaimed lakes have a characteristic grid-like layout.

reed peat *rietveen* **p. 22, 23 (15B), 25 (16), 26 (17)** Peat consisting largely of reed remains. Reed peat often contains some clay deposited by river or sea flooding.

reed-sedge peat *riet-zeggenveen* **p. 25, 26 (17)** Peat composed of the remains of reeds and sedge.

regression *regressie* **p. 17** In geology, the retreat of the sea from the land.

river dune *rivierduin* **p. 21 (13), 33** A **dune** created along the big rivers at the end of the Pleistocene and the beginning of the Holocene when sand was blown from the non-submerged parts of braided rivers.

Romanisation *romanisering* **pp. 66-67** The process of social and cultural change in non-Roman societies as a result of contact (at times violent, but more often peaceful) with Roman culture, Roman society and the Roman state.

salt extraction from peat *moernering or selnering* **p. 66, 74** The extraction of salt from peat that is saturated in sea water. This involved digging up, drying and then burning the peat, a process in which the salt was left behind. For a long time this was the main source of salt in the Netherlands. This salt-extraction method was used on a large scale in Zeeland and was one of the causes of major flooding. Large tracts of land were lost in this way during the Middle Ages.

salt marsh *kwelder or schor* **p. 41, 45, 49 (39)** The supratidal area at the margins of tidal basins and estuaries. Salt marshes are only flooded at spring tide and stormfloods, causing marine clay to be deposited. Because of this occasional flooding, a salt-tolerant vegetation grows in salt marshes.

salt-marsh clay layer *kwelderkleidek* **p. 70** A surface salt-marsh clay layer.

sandhill *zandkop* See coversand hill.

sea-level rise *absolute zeespiegelstijging* **pp. 13-16 (7), (8), (9)** Rise in sea level as a result of increased water volumes in the oceans.

sea-level rise, relative *relatieve zeespiegelstijging* **p. 15, 20, 27, 44, 56, 63** The rise in sea level as the net result of absolute sea-level fluctuations and factors at a more local level such as sedimentation, erosion and ground movements.

sedge peat *zeggenveen* **pp. 22-26 (15C-17)** Peat that is mainly composed of the remains of sedge plants.

(clastic) sediment **passim** Eroded material transported by water, wind or ice sheet. Gravel, sand, clay and loess are all sediments, whereas peat, which is formed in situ from plant remains and water, is not.

single grave culture *enkelgrafcultuur* **p. 51** A culture of groups from the late Neolithic whose pan-regional burial custom involved interring a single body in a grave, in contrast to the collective burials in the chamber tombs (or *hunebedden*) of the Funnelbeaker culture. In the Netherlands and northwest Germany the single grave culture is the northwestern branch of what are called the Battle Axe or Corded Ware cultures (*Schnurkeramik*), which occupied large parts of Europe. They are named respectively after the commonly found stone battle axes (or, more neutrally, 'hammer axes') and beaker-shaped pots decorated with cord.

soil compaction/consolidation *klink* **pp. 15-16** The sinking of unconsolidated soil layers (clay and peat) as a result of dehydration and **oxidation**.

Sphagnum peat *veenmosveen* **pp. 24-26 (15E-17)** Peat consisting of the remains of *Sphagnum* moss.

spreng (pl. sprengen) **p. 83** A *spreng* is a dug water course or realigned stream that has been dug back in such a way as to bring pressurised ground water to the surface. *Sprengen* are extensions of a natural stream valley: digging back closer to the source allowed more water into the stream. *Sprengen* occur mainly on and around the Veluwe, where they once played a key role in the now-vanished paper industry. Nowadays *sprengen* are maintained for their natural and heritage values.

spring tide *springtij* **p. 56 (44)** Tidal cycle with the greatest difference between high and low tide. It occurs when the Sun, Moon and Earth are aligned and the tidal-producing forces of the sun and moon reinforce one another. This happens at around the time of the new moon and full moon (i.e. about once every two weeks). Neap tide, the tidal cycle with the smallest tidal range, occurs when the Sun and Moon are perpendicular in relation to the Earth and therefore their tidal-producing forces do not supplement each other. Neap tide also occurs twice a month, during the Moon's first and last quarters.

steppes *steppe* **p. 43** Virtually treeless plains with a grassland vegetation. Steppes develop in areas that are dry for at least eight months of the year, as was often the case in the Netherlands in the cold parts of the Pleistocene.

storm surge *stormvloed* **p. 56 (44), 78, 84** Storm surges are caused by high winds driving the sea water. If the storm occurs at high tide, the water level can be several metres higher than normal. Storm surges that coincide with a spring tide can endanger the dykes. In the Netherlands this happens when the wind comes from the north-west, driving the water from the North Sea into the narrow funnel that is the English Channel. A storm surge of this type led to the catastrophic North Sea flood in the southwestern Netherlands in 1953, when 1836 people drowned.

subsidence *bodemdaling* **pp. 15-17 (9)** The sinking of the Earth's surface. This can be due to tectonic processes (see **tectonic subsidence**), the after-effects of ice load, or drainage, whereby layers of peat and clay degrade or subside (see also **sea-level rise**).

subsidence, tectonic *tektonische bodemdaling* **p. 15** The sinking of the Earth's crust, often along faults.

subtidal landscape *subgetijdenlandschap* **p. 18** The area located below mean low water. It includes large tidal channels, sand banks and tidal inlets.

supratidal zone *supragetijdengebied* **p. 18** The area that lies above mean high water. It is only inundated during spring tide and major storms.

Swifterbant culture *Swifterbantcultuur* **p. 47** A culture from the Middle Neolithic (c. 5000-3400 BCE), during the transition from a hunter-gatherer way of life to one based on agriculture. It took some 1500 years before the inhabitants of the Dutch delta fully abandoned the traditional lifestyle dominated by hunting, fishing and gathering and settled in small, permanent farming communities.

terp (or wierde or woerd) (pl. terpen) p. 56, 59 An artificial dwelling mound for one or more farmhouses, composed of sods cut from the salt marshes, clay, manure or waste. *Terpen* or *wierden* occur in the salt-marsh areas of Friesland and Groningen respectively, while *woerden* are typically found in river areas.

tidal basin *getijdenbekken* **p. 14, (11), 19 (12), (16), 42, 44, 48** A very shallow part of the sea that is separated from the open sea by beach barriers. At every high tide the sea water flows through tidal inlets between the beach barriers into the tidal basin, filling it with water. At low tide the sea water flows back through the tidal inlets into the open sea, leaving large parts of the tidal basin dry (see **mudflats**). Small local rivers from the hinterland can find their way to the sea through a tidal basin.

tidal channel *getijdengeul* **p. 18, (39), 63 (520 (55)** A channel in a tidal basin through which water flows at high and low tide. Tidal channels are not dry at low tide.

tidal flats or sand- and mudflats *wad* **p. 25 (16), 33** Part of a tidal basin that is submerged at high tide and falls dry at low tide. Tidal flats consist of sand- and mudflats or shoals.

tidal inlet *zeegat* **p. 18 (11), 52, 56** An opening between two beach barriers or islands that links the open sea in front of them with the tidal basin behind them. The **tidal volume** flows through the inlet into the tidal basin twice a day (at high tide) and flows out again twice a day (at low tide). This means that the current in the tidal inlet is often very strong, causing deep scouring of the inlet.

tidal landscape *getijdenlandschap* **p. 18** The landscape in a tidal basin, made up of salt marshes, tidal flats or sand- and mudflats and tidal channels.

tidal system *getijdensysteem* **p. 44, 60** Holistic, generic term for the tidal landscape and its components, the hydrodynamic and sedimentological processes that occur there and the resulting deposits. Geologists use the term both to describe current systems and to interpret fossil systems which, although no longer observable, may be deduced from deposition characteristics. This atlas describes two tidal systems for the Netherlands: **tidal basins** and **estuaries**.

tidal volume *getijdenvolume* **p. 19** The amount of water flowing through a tidal inlet into a tidal basin at high tide and flowing out again at low tide.

till *keileem* **p. 33** Sandy loam mixed with pebbles and stone debris that has been transported by land ice, formed beneath a shifting ice cap.

till outcrop *keileemopduiking* **p. 33** Low hill composed of boulder clay that has been pushed up by an ice sheet.

tow-canal *trekvaart* **p. 83** A tow-canal is a waterway especially dug for horse-drawn boats (*trekschuiten*). The boats were pulled along by a horse or person on a towpath (*jaagpad* in Dutch, named after the *jager* – or 'chaser'– who led the horse), which ran beside the canal. The advantage of tow-canals was that they shortened the transport route between destinations. Tow-canals also enabled the regular transportation of people and mail because there was no longer a reliance on the wind. In the nineteenth century, transport by tow-canal was overtaken by rail transport.

transgression *transgressie* **p. 25** In geology, the sea advancing across a former land area.

villa p. 67 A Roman-period farming estate whose products were destined primarily for the market. It also served as a showcase for the owner's consumer lifestyle, which could be expressed in the sumptuous interiors and exteriors of the main building and bathhouse. In the Netherlands, villas are mainly found on the fertile loess soils and in areas that had good road links.

Vlaardingen culture *Vlaardingencultuur* **p. 50** A culture from the last part of the Middle Neolithic (c. 3400-2500 BCE). They were farmers who lived in the more elevated parts of the western Netherlands. Hunting and fishing continued to play a key role.

wierde See **terp**

woerd See **terp**

wood peat *bosveen* **p. 25 (16)** Peat containing the remains of trees, usually birch and alder. Where the wood peat directly overlies coversand, there are often pine tree stumps (Dutch *stobben*) at the base of the peat.

wood tar *houtteer* **p. (31), 43** Wood tar is extracted by heating wood in a dry, sealed space. In prehistory it was used to fasten flint arrowheads to arrow shafts. It was also used as a lubricant or preservative for wood, rope, etc.

ACKNOWLEDGEMENTS

Authors

Chapter 1: Jos Bazelmans, Bob Hoogendoorn, Michiel van der Meulen, Peter Vos and Henk Weerts

Chapter 2: Henk Weerts, Bob Hoogendoorn, Michiel van der Meulen and Peter Vos

Chapter 3: Peter Vos, Henk Weerts, Michiel van der Meulen and Bob Hoogendoorn

Chapter 4: Peter Vos and Jos Bazelmans

Chapter 5: Henk Weerts, Peter Vos, Bob Hoogendoorn and Michiel van der Meulen

Chapter 6: Henk Weerts, Peter Vos, Bob Hoogendoorn and Michiel van der Meulen

Chapter 7: Jos Bazelmans

Chapter 8: Peter Vos, Jos Bazelmans

Chapter 9: Peter Vos, Henk Weerts

Text accompanying map for 9000 BCE: Eelco Rensink, Peter Vos, Henk Weerts, Roel Lauwerier and Otto Brinkkemper

Text accompanying map for 5500 BCE: Hans Peeters, Peter Vos, Henk Weerts, Roel Lauwerier and Otto Brinkkemper

Text accompanying map for 3850 BCE: Hans Peeters, Peter Vos, Henk Weerts, Roel Lauwerier and Otto Brinkkemper

Text accompanying map for 2750 BCE: Liesbeth Theunissen, Peter Vos, Henk Weerts, Roel Lauwerier and Otto Brinkkemper

Text accompanying map for 1500 BCE: Stijn Arnoldussen, Peter Vos, Henk Weerts, Roel Lauwerier and Otto Brinkkemper

Text accompanying map for 500 BCE: Liesbeth Theunissen, Peter Vos, Henk Weerts, Jos Bazelmans, Roel Lauwerier and Otto Brinkkemper

Text accompanying map for 250 BCE: Peter Vos and Liesbeth Theunissen

Text accompanying map for 100 CE: Tessa de Groot, Peter Vos, Henk Weerts, Jan Willem de Kort, Jan van Doesburg, Roel Lauwerier and Otto Brinkkemper

Text accompanying map for 800 CE: Jan van Doesburg, Peter Vos, Henk Weerts, Bert Groenewoudt, Roel Lauwerier and Otto Brinkkemper

Text accompanying map for 1250 CE: Peter Vos, Jos Bazelmans en Michiel Lascaris

Text accompanying map for 1500 CE: Bert Groenewoudt, Peter Vos, Henk Weerts, Theo Spek, Roel Lauwerier and Otto Brinkkemper

Text accompanying map for 1850 CE: Lammert Prins, Peter Vos, Henk Weerts, Roel Lauwerier and Otto Brinkkemper

Text accompanying map for 2000 CE: Henk Weerts, Henk Baas, Peter Vos, Roel Lauwerier and Otto Brinkkemper

Glossary: Henk Weerts and Jos Bazelmans

Glossary

The glossary is based on the following sources:

- J. Deeben, E. Drenth, M.-F. van Oorsouw and L. Verhart (eds), *De steentijd van Nederland (Archeologie 11/12)* (Meppel 2005).
- L.P. Louwe Kooijmans, P.W. van de Broeke, H. Fokkens and A.L. van Gijn, *Nederland in de prehistorie* (Amsterdam 2005).
- M. van der Meulen, F.D. de Lang, D. Maljers, C.W. Dubelaar and W.E. Westerhoff, *Grondsoorten en delfstoffen bij naam* (Publicatiereeks grondstoffen 2003/16) (Delft 2003, second, slightly amended edition).
- W.A. Visser (ed.), *Geological nomenclature* (Utrecht/Den Haag 1980).
- J.B. Whittow, *The Penguin dictionary of physical geography* (London 1984).
- www.geologievannederland.nl.

Figures

Fig. 1 From W.H. Zagwijn, *Nederland in het Holoceen* (Geologie van Nederland 1) ('s-Gravenhage 1986).

Fig. 2 After: E.F.J. de Mulder, M.C. Geluk, I. Ritsema, W.E. Westerhoff and Th.E. Wong, *De ondergrond van Nederland* (Geologie van Nederland 7) (Utrecht 2003).

Fig. 3 After: H.J.A. Berendsen, *De vorming van het land. Inleiding in de geologie en de geomorfologie* (Assen 2004; fourth completely revised edition); J.R. Petit, J. Jouzel, D. Raynaud, N.I. Barkov, J.-M. Barnola, I. Basile, M. Bender, J. Chappellaz, M. Davis, G. Delaygue, M. Delmotte, V.M. Kotlyakov, M. Legrand, V.Y. Lipenkov, C. Lorius, L. Pépin, C. Ritz, E. Saltzman and M. Stievenard, 'Climate and atmospheric history of the past 420,000 years from the Vostok ice core, Antarctica', *Nature* 399 (1999) 429-436.

Fig. 4 After: J. Bazelmans et al.: Oud hout uit een Droogmakerij. Laat-glaciale bosresten uit Leusden-Den Treek, in J. Bazelmans and L. Voets (eds), *Jaarboek Nederlandse Archeologie* 2017 (Vught 2018), 25-30.

Fig. 5 After: V. Gaffney, K. Thomson and S. Fitch (eds), *Mapping Doggerland. The Mesolithic landscapes of the Southern North Sea* (Birmingham 2007).

Fig. 6 After: F.S. Busschers, C. Kasse, R.T. van Balen, J. Vandenberghe, K.M. Cohen, H.J.T. Weerts, J. Wallinga, C. Johns, P. Cleveringa and F.P.M. Bunnik, 'Late Pleistocene evolution of the Rhine-Meuse system in the southern North Sea basin. Imprints of climate change, sea-level oscillation and glacio-isostacy', *Quaternary Science Reviews* 26 (2007) 3216-3248; H. Steffen, *Determination of a consistent viscosity distribution in the Earth's mantle beneath Northern and Central Europe* (Berlin 2006).

Fig. 7 After: J.H.J. Ebbing, H.J.T. Weerts and W.E. Westerhoff, 'Towards an integrated land-sea stratigraphy of the Netherlands', *Quaternary Science Reviews* 22 (2003) 1579-1587; H.J.A. Berendsen, *De vorming van het land. Inleiding in de geologie en de geomorfologie* (Assen 1996); J. Cleveringa, 'Reconstruction and modelling of Holocene coastal evolution of the western Netherlands', *Geologica Ultraiectina/Mededelingen van de Faculteit Aardwetenschappen Universiteit Utrecht* 200 (2000) 1-198.

Fig. 8 After: P. Kiden, B. Makaske and O. van de Plassche, 'Waarom verschillende zeespiegelreconstructies voor Nederland?', *Grondboor en Hamer* 62 (2008) 54-61; A. Vink, et al. Holocene relative sea-level change, isostatic subsidence and the radial viscosity structure of the mantle of Northwest Europe (Belgium, the Netherlands, Germany, southern North Sea)', *Quaternary Science Reviews* 26 (2007) 3249-3275.

Fig. 9 After: H.J.A. Berendsen, *De vorming van het land. Inleiding in de geologie en de geomorfologie* (Assen 2004; fourth completely revised edition); W.F. Ruddiman, *Earth's climate. Past and future* (New York 2001).

Fig. 10 After: J. Houbolt, Recent sediments in the Southern Bight of the North Sea, *Geologie en Mijnbouw* 1968, 248 (figure 3); Martinius and Van den Berg, Atlas of sedimentary structures in estuarine and tidally-influenced river deposits of the Rhine-Meuse-Scheldt system. Their application to the interpretation of analogous outcrop and subsurface depositional systems (EAGE publication, Houten, 2011) and *Wadatlas Rijkswaterstaat* (s.l. 1989).

Fig. 11 After: P.C. Vos and Y. Eijskoot, *Geo- en archeolandschappelijk onderzoek bij de opgravingen van de Vergulde Hand West (VHW) in Vlaardingen* (Deltares rapport 0912-0245) (Utrecht 2009); H.-E. Reineck, 1970. 'Das Watt als Ablagerungsbereich', in: H.-E. Reineck (ed.), *Das Watt. Ablagerungs- und Lebensraum* (Frankfurt am Main 1970) 131-135; H.G. Reading and J.D. Collinson, 'Clastic coasts', in: H.G. Reading (ed.), *Sedimentary environments. Processes, facies and stratigraphy* (Oxford 1996) 154-231.

Fig. 12 After: P.C. Vos, *Origin of the Dutch coastal landscape. Long-term landscape evolution of the Netherlands during the Holocene* (Groningen 2015) 46 (figure 1.30).

Fig. 13 After: H.J.T. Weerts, 'Complex confining layers. Architecture and hydraulic properties of Holocene and Late Weichselian deposits in the fluvial Rhine-Meuse delta, The Netherlands', *Nederlandse Geografische Studies* 213 (1996) 1-189; A.D. Miall, 'Architectural-element analysis. A new method of facies analysis applied to fluvial deposits', *Earth-Science Reviews* 22 (1985) 261-308; H.J.A. Berendsen and E. Stouthamer, *Palaeogeographic development of the Rhine-Meuse delta, The Netherlands* (Assen 2001).

Fig. 14 Jerry Struik, Amsterdam.

Fig. 15a Paul Paris Les Images, Amstelveen.

Fig. 15b and c Laura I. Kooistra (BIAX Consult).

Fig. 15d Lex Broere Buitenfotografie, Nieuwerkerk a/d IJssel.

Fig. 15e Paul Paris Les Images, Amstelveen.

Fig. 16 After: P.C. Vos and Y. Eijskoot, *Geo- en archeolandschappelijk onderzoek bij de opgravingen van de Vergulde Hand West (VHW) in Vlaardingen* (Deltares rapport 0912-0245) (Utrecht 2009); H.G. Reading & J.D. Collinson, 'Clastic coasts', in: H.G. Reading (ed.), *Sedimentary environments.*

Processes, facies and stratigraphy (Oxford 1996) 154-231; H.E. Reineck, 'Das Watt als Ablagerungsbereich', in: H.-E. Reineck (ed.), *Das Watt. Ablagerungs- und Lebensraum* (Frankfurt am Main 1970) 131-135; M. van Dinter, *Living along the Limes; Landscape and settlement in the Lower Rhine Delta during Roman and Early Medieval times* (Utrecht 2017).

Fig. 17 After: P.C. Vos, *Origin of the Dutch coastal landscape. Long-term landscape evolution of the Netherlands during the Holocene* (Groningen 2015) 33 (figure 1.17).

Fig. 18 H.A. Hiddink, Vrije Universiteit Amsterdam.

Fig. 19 B. Maes, 2013: Inheemse bomen en struiken in Nederland en Vlaanderen. Herkenning, verspreiding, geschiedenis en gebruik (Amsterdam, 2013), 36. The figure is by O. Brinkkemper, with assistance from Jan Bastiaers and Koen Deforce.

Fig. 20 Universiteitsbibliotheek Utrecht.

Fig. 21 Deltares.

Fig. 22a Actueel Hoogtebestand Nederland.

Fig. 22b After: P.C. Vos, *Origin of the Dutch coastal landscape. Long-term landscape evolution of the Netherlands during the Holocene* (Groningen 2015) 114 (figure 3.2.6f).

Fig. 23 After: P.C. Vos, *Origin of the Dutch coastal landscape. Long-term landscape evolution of the Netherlands during the Holocene* (Groningen 2015) 284 (figures 4.2.13d and e).

Fig. 24 National Park Services Canada, Neal Herbert.

Fig. 25 Rijksmuseum van Oudheden, Leiden.

Fig. 26 TGV Teksten & Presentaties, Yuri van Koeveringe.

Fig. 27 After: M. Bats, Ph. Crombé, I. Devriendt, R. Langohr, J.H. Mikkelsen, C. Ryssaert and A. Van de Water, *Een vroegmesolithische vindplaats te Haelen-Broekweg (gem. Leudal, provincie Limburg).* Archeologie in de A73-Zuid; Rapportage Archeologische Monumentenzorg 190 (Amersfoort 2010).

Fig. 28 Ralf Lotys.

Fig. 29 Rijksmuseum van Oudheden, Leiden.

Fig. 30 TGV Teksten & Presentaties, Yuri van Koeveringe.

Fig. 31 After: L. Kubiak-Martens, L.I. Kooistra and J.J. Langer, 'Mesolithische teerproductie in Hattemerbroek', in: E. Lohof, T. Hamburg and J. Flamman, *Steentijd opgespoord. Archeologisch onderzoek in het tracé van de Hanzelijn-Oude Land* (Archolrapport 138/ADC-rapport 2576) (Leiden/ Amersfoort 2011), 491-506.

Fig. 32 Vildaphoto, Yves Adams.

Fig. 33 Universiteit Leiden, Faculteit Archeologie.

Fig. 34 P.J.R. Modderman, *Linearbandkeramik aus Elsloo und Stein* (Leiden 1970) 91 (Fig. 7).

Fig. 35 After: P.H. Deckers, J.P. de Roever and J.D. van der Waals, 'Jagers, vissers en boeren in een prehistorisch getijdengebied bij Swifterbant', *ZWO jaarboek* (1980) 111-144.

Fig. 36 Henk Weerts.

Fig. 37 Frans de Vries.

Fig. 38 Frans de Vries.

Fig. 39 After: the palaeogeographic map of 2750 BCE in this publication and J.M. Pasveer; H.T. Uytterschat, 'Twee laat-neolithische skeletten uit Noord-Holland. Een fysisch-antropologisch onderzoek', *Westerheem* 41 (1992) 268-275.

Fig. 40 Henk Weerts.

Fig. 41 Rijksmuseum van Oudheden, Leiden.

Fig. 42 Frans de Vries.

Fig. 43 After: S. Arnoldussen, *A living landscape. Bronze Age settlement sites in the Dutch River area (c. 2000-800 BC)* (Leiden 2008).

Fig. 44 Bart van der Schacht.

Fig. 45 Gemeente Rotterdam, Bureau Oudheidkundig Onderzoek.

Fig. 46 Rijksuniversiteit Groningen.

Fig. 47 After: P. Kooi and K. van der Ploeg, *Ezinge. IJkpunt in de archeologie* (Leiden, 2014); Actueel Hoogstebestand Nederland.

Fig. 48 Dick Bos.

Fig. 49 Vrije Universiteit Amsterdam.

Fig. 50 Frans de Vries.

Fig. 51 After: J. Bazelmans, D. Gerrets, J. de Koning and P. Vos, 'Zoden aan de dijk. Kleinschalige dijkbouw in de late prehistorie en protohistorie van noordelijk Westergo', *De Vrije Fries* 79 (1999) 7-74.

Fig. 52 Szabo Szilard, Panoramio.

Fig. 53 Cultural Heritage Agency of the Netherlands (RCE).

Fig. 54 TGV Teksten & Presentaties, Evert van Ginkel.

Fig. 55 After: J.W. de Kort, 'Het kanaal van Corbulo', in: R.L. Hirschel (ed.), *Forum Hadriani. Romeinse stad achter de Limes* (Voorburg 2009) 24-28; B. Jansen, *De bocht in kanaal van Corbulo. Resultaten proefsleuvenonderzoek plangebied Leytsehif, gemeente Leidschendam-Voorburg* (in prep., Weesp).

Fig. 56 Yves Adams, Vilda.

Fig. 57 Cultural Heritage Agency of the Netherlands (RCE).

Fig. 58 Erfgoedfoto, Jos Stöver.

Fig. 59 After: H.J.A. Berendsen, *De genese van het landschap in het zuiden van de provincie Utrecht. Een fysisch-geografische studie* (Utrechtse Geografische Studies 10) (Utrecht 1982); K.M. Cohen, E. Stouthamer, W.Z. Hoek, H.J.A. Berendsen and H.F.J. Kempen, *Zand in Banen. Zanddieptekaarten van het rivierengebied en het IJsseldal in de provincies Gelderland en Overijssel* (Arnhem 2009; third completely revised edition); W.A. van Es, 'Friezen, Franken en Vikingen', in: W.A. van Es and W.A.M. Hessing (eds), *Romeinen, Friezen en Franken in het hart van Nederland. Van Trajectum tot Dorestad 50 BCE-900 n.Chr* (Utrecht/Amersfoort 1994) 82-119.

Fig. 60 Erfgoedfoto, Jos Stöver.

Fig. 61 Gemeente Rotterdam, Bureau Oudheidkundig Bodemonderzoek.

Fig. 62 Cultural Heritage Agency of the Netherlands (RCE).

Fig. 63 After: M.H. Bartels (ed.), *Dwars door de dijk. Archeologie en geschiedenis van de Westfriese Omringdijk tussen Hoorn en Enkhuizen* (Hoorn 2016); M. Bartels, De Westfriese Omringdijk Archeologisch en historisch onderzoek tussen Hoorn en Enkhuizen, *Archeologie in Nederland* (Utrecht 2017), 34-41.

Fig. 64 Paul Paris Les Images, Amstelveen.

Fig. 65 Rijksmuseum Amsterdam.

Fig. 66 Cultural Heritage Agency of the Netherlands (RCE).

Fig. 67 After: B.J. Groenewoudt, 'Sporen van oud groen. Bomen en bos in het historische cultuurlandschap van Zutphen-Looërenk', in O. Brinkkemper, J. Deeben, J. van Doesburg, D.P. Hallewas, E.M.Theunissen and A.D.Verlinde (eds), *Vlakken in vakken. Archeologische kennis in lagen* (Nederlandse Archeologische Rapporten 32) (Amersfoort 2006) 117-146.

Fig. 68 Siebe Swart.

Fig. 69 Fred Oudejans.

Fig. 70 Cultural Heritage Agency of the Netherlands (RCE).

Fig. 71 After: Bonneblad 279 (1879 edition); Bonneblad 280 (1879 edition).

Fig. 72 Aerophoto-Schiphol, Marco van Middelkoop.

Fig. 73 Havenbedrijf Rotterdam, Projectbureau Uitvoering Maasvlakte, Rogier Scholten.

Fig. 74 Erfgoedfoto, Jos Stöver.

Fig. 75 After information from Havenbedrijf Rotterdam, 2010.

5500 BCE

3850 BCE

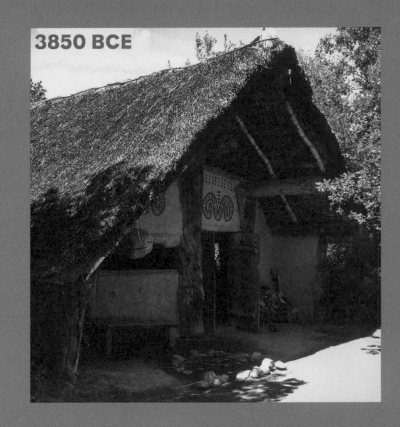

500 BCE

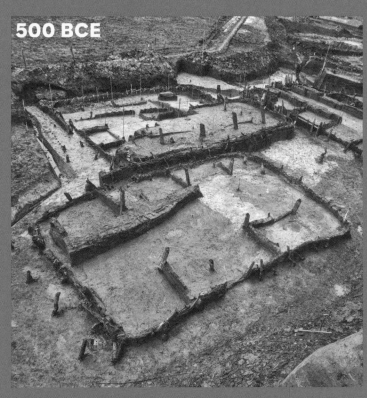

250 BCE

1250 CE

1500 CE